U.S.NRC

United States Nuclear Regulatory Commission

Protecting People and the Environment

NUREG-1920, Vol. 2

I0482684

Safety Evaluation Report

Related to the License Renewal of Vogtle Electric Generating Plant, Units 1 and 2

Docket Nos. 50-424 and 50-425

Southern Nuclear Operating Company, Inc.

Office of Nuclear Reactor Regulation

AVAILABILITY OF REFERENCE MATERIALS
IN NRC PUBLICATIONS

NRC Reference Material

As of November 1999, you may electronically access NUREG-series publications and other NRC records at NRC's Public Electronic Reading Room at http://www.nrc.gov/reading-rm.html.
Publicly released records include, to name a few, NUREG-series publications; *Federal Register* notices; applicant, licensee, and vendor documents and correspondence; NRC correspondence and internal memoranda; bulletins and information notices; inspection and investigative reports; licensee event reports; and Commission papers and their attachments.

NRC publications in the NUREG series, NRC regulations, and *Title 10, Energy,* in the Code of *Federal Regulations* may also be purchased from one of these two sources.
1. The Superintendent of Documents
   U.S. Government Printing Office
   Mail Stop SSOP
   Washington, DC 20402-0001
   Internet: bookstore.gpo.gov
   Telephone: 202-512-1800
   Fax: 202-512-2250
2. The National Technical Information Service
   Springfield, VA 22161-0002
   www.ntis.gov
   1-800-553-6847 or, locally, 703-605-6000

A single copy of each NRC draft report for comment is available free, to the extent of supply, upon written request as follows:
Address: U.S. Nuclear Regulatory Commission
           Office of Administration
           Reproduction and Mail Services Branch
           Washington, DC 20555-0001
E-mail: DISTRIBUTION@nrc.gov
Facsimile: 301-415-2289

Some publications in the NUREG series that are posted at NRC's Web site address
http://www.nrc.gov/reading-rm/doc-collections/nuregs
are updated periodically and may differ from the last printed version. Although references to material found on a Web site bear the date the material was accessed, the material available on the date cited may subsequently be removed from the site.

Non-NRC Reference Material

Documents available from public and special technical libraries include all open literature items, such as books, journal articles, and transactions, *Federal Register* notices, Federal and State legislation, and congressional reports. Such documents as theses, dissertations, foreign reports and translations, and non-NRC conference proceedings may be purchased from their sponsoring organization.

Copies of industry codes and standards used in a substantive manner in the NRC regulatory process are maintained at—
           The NRC Technical Library
           Two White Flint North
           11545 Rockville Pike
           Rockville, MD 20852-2738

These standards are available in the library for reference use by the public. Codes and standards are usually copyrighted and may be purchased from the originating organization or, if they are American National Standards, from—
           American National Standards Institute
           11 West 42nd Street
           New York, NY 10036-8002
           www.ansi.org
           212-642-4900

Legally binding regulatory requirements are stated only in laws; NRC regulations; licenses, including technical specifications; or orders, not in NUREG-series publications. The views expressed in contractor-prepared publications in this series are not necessarily those of the NRC.

The NUREG series comprises (1) technical and administrative reports and books prepared by the staff (NUREG-XXXX) or agency contractors (NUREG/CR-XXXX), (2) proceedings of conferences (NUREG/CP-XXXX), (3) reports resulting from international agreements (NUREG/IA-XXXX), (4) brochures (NUREG/BR-XXXX), and (5) compilations of legal decisions and orders of the Commission and Atomic and Safety Licensing Boards and of Directors' decisions under Section 2.206 of NRC's regulations (NUREG-0750).

United States Nuclear Regulatory Commission

Protecting People and the Environment

NUREG-1920, Vol. 2

Safety Evaluation Report

Related to the License Renewal of Vogtle Electric Generating Plant, Units 1 and 2

Docket Nos. 50-424 and 50-425

Southern Nuclear Operating Company, Inc.

Manuscript Completed: April 2009
Date Published: April 2009

Office of Nuclear Reactor Regulation

ABSTRACT

This safety evaluation report (SER) documents the technical review of the Vogtle Electric Generating Plant (VEGP), Units 1 and 2, license renewal application (LRA) by the United States (US) Nuclear Regulatory Commission (NRC) staff (the staff). By letter dated June 29, 2007, Southern Nuclear Operating Company, Inc. (SNC or the applicant) submitted the LRA in accordance with Title 10, Part 54, of the *Code of Federal Regulations*, "Requirements for Renewal of Operating Licenses for Nuclear Power Plants." SNC requests renewal of the Units 1 and 2 operating licenses (Facility Operating License Numbers NPF-68 and NPF-81, respectively) for a period of 20 years beyond the current expiration date of January 16, 2027, for Unit 1, and February 9, 2029, for Unit 2.

VEGP is located approximately 26 miles southeast of Augusta, GA. The NRC issued the construction permits for Unit 1 on June 28, 1974, and on June 28, 1974, for Unit 2. The NRC issued the operating licenses for Unit 1 on March 16, 1987, and on March 31, 1989, for Unit 2. Units 1 and 2 are of a dry ambient containment pressurized water reactor design. Westinghouse Electric supplied the nuclear steam supply system and Georgia Power Company originally designed and constructed the balance of the plant with the assistance of its agent, Southern Services and Bechtel. The licensed power output of each unit is 3625.6 megawatt thermal with a gross electrical output of approximately 1250 megawatt electric.

This SER presents the status of the staff's review of information submitted through February 16, 2009, the cutoff date for consideration in the SER. The staff identified no open or confirmatory items that would require a formal response from the applicant. SER Section 6 provides the staff's final conclusion of its LRA review.

TABLE OF CONTENTS

Tables

ABBREVIATIONS

AB	auxiliary building
ACI	American Concrete Institute
ACRS	Advisory Committee on Reactor Safeguards
ADAMS	Agencywide Document Access and Management System
AERM	aging effect requiring management
AFW	auxiliary feedwater
AISC	American Institute of Steel Construction
AMP	aging management program
AMR	aging management review
AMSAC	ATWS mitigation system actuation circuitry
ANSI	American National Standards Institute
ART	adjusted reference temperature
ASME	American Society of Mechanical Engineers
ASTM	American Society for Testing and Materials
ATWS	anticipated transient without scram
AWWA	American Water Works Association
BAC	boric acid corrosion
BWR	boiling water reactor
B&PV	boiler and pressure vessel
CASS	cast austenitic stainless steel
CCW	component cooling water
CET	core exit thermocouple
CF	chemistry factor
CFR	*Code of Federal Regulations*
CI	confirmatory item
CLB	current licensing basis
COPS	cold overpressure protection system
CRDM	control rod drive mechanism
CR	condition report
CRGT	control rod guide tube
CS	containment spray
CST	condensate storage tank
CTMT	containment
CTB	containment building
CUF	cumulative usage factor
CVCS	chemical and volume control system
DAW	dry active waste
DBA	design basis accident
DBE	design basis event
DC	direct current
DW	demineralized water
ECCS	emergency core cooling system
EDG	emergency diesel generator
EFPY	effective full-power year

EHC	electrohydraulic control
EOL	end of life
EPRI	Electric Power Research Institute
EQ	environmental qualification
ESF	engineered safety feature
FAC	flow-accelerated corrosion
F_{en}	environmental fatigue life correction factor
FP	fire protection
FPP	fire protection plan
FR	*Federal Register*
FRRADS	flood-retaining rooms, alarms, and drain system
FW	feedwater
GALL	Generic Aging Lessons Learned Report
GDC	general design criteria or general design criterion
GEIS	Generic Environmental Impact Statement
GL	generic letter
GPC	Georgia Power Company
GSI	generic safety issue
HAZ	heat-affected zone
HELB	high-energy line break
HE/ME	high energy/moderate energy
HJTC	heated junction thermocouple
HVAC	heating, ventilation, and air conditioning
HX	heat exchanger
I&C	instrumentation and controls
IASCC	irradiation assisted stress corrosion cracking
IEEE	Institute of Electrical and Electronics Engineers
IGA	intergranular attack
IN	information notice
INPO	Institute of Nuclear Power Operations
IPA	integrated plant assessment
ISG	interim staff guidance
ISO	International Organization for Standardization
ISI	inservice inspection
LBB	leak-before-break
LOCA	loss of coolant accident
LOSP	loss of offsite power
LR	license renewal
LRA	license renewal application
MPL	master parts list
MSLB	main steam line break
MWe	megawatts electric
MWt	megawatts thermal
NDE	nondestructive examination

NEI	Nuclear Energy Institute
NPS	nominal pipe size (in inches)
NRC	U.S. Nuclear Regulatory Commission
NSCW	nuclear service cooling water
NSR	nonsafety-related
NSSS	nuclear steam supply system
OBE	operating basis earthquake
ODSCC	outside-diameter stress corrosion cracking
OI	open item
P&ID	piping and instrumentation diagram
PRF	penetration room filtration
PSRF	nonsafety-related that can prevent a safety-related function
PTLR	pressure-temperature limits report
PTS	pressurized thermal shock
PVC	polyvinyl chloride
PW	pipe whip
PWR	pressurized water reactor
PWSCC	primary water stress corrosion cracking
RAI	request for additional information
RCP	reactor coolant pump
RCPB	reactor coolant pressure boundary
RCS	reactor coolant system
RG	regulatory guide
RHR	residual heat removal
RI-ISI	risk-informed inservice inspection
RMWST	reactor makeup water storage tank
RPV	reactor pressure vessel
RT_{NDT}	reference temperature for nil ductility transition
RT_{PTS}	reference temperature for pressurized thermal shock
RTS	reactor trip system
RVCH	reactor vessel closure head
RVLIS	reactor vessel level indicating system
RWST	refueling water storage tank
RV	reactor vessel
SBO	station blackout
SCs	structures and components
SCC	stress-corrosion cracking
SER	safety evaluation report
SFP	spent fuel pool
SG	steam generator
SGBD	steam generator blowdown
SI	safety injection
SMP	structural monitoring program
SNC	Southern Nuclear Operating Company, Inc.
SOC	statement of consideration
SR	safety-related
SRP	Standard Review Plan

SRP-LR	Standard Review Plan for Review of License Renewal Applications for Nuclear Power Plants
SSCs	systems, structures, and components
SSE	safe-shutdown earthquake
SW	service water
TLAA	time-limited aging analysis
TS	technical specifications
TSP	trisodium phosphate
UFSAR	updated final safety analysis report
USE	upper-shelf energy
UT	ultrasonic testing
UV	ultraviolet
VEGP	Vogtle Electric Generating Plant
WCAP	Westinghouse Commercial Atomic Power
WOG	Westinghouse Owner's Group

SECTION 4

TIME-LIMITED AGING ANALYSES

4.1 Identification of Time-Limited Aging Analyses

This section of the safety evaluation report (SER) addresses the identification of time-limited aging analyses (TLAAs). In license renewal application (LRA) Sections 4.2 through 4.7, Southern Nuclear Operating Company, Inc. (SNC or the applicant) addressed the TLAAs for Vogtle Electric Generating Plant (VEGP), Units 1 and 2. SER Sections 4.2 through 4.8 documents the review of the TLAAs conducted by the staff of the United States Nuclear Regulatory Commission (NRC) (the staff).

TLAAs are certain plant-specific safety analyses that involve time-limited assumptions defined by the current operating term. Pursuant to Title 10, Section 54.21(c)(1), of the *Code of Federal Regulations* (10 CFR 54.21(c)(1)), applicants must list TLAAs as defined in 10 CFR 54.3.

In addition, pursuant to 10 CFR 54.21(c)(2), applicants must list plant-specific exemptions granted under 10 CFR 50.12 based on TLAAs. For any such exemptions, the applicant must evaluate and justify the continuation of the exemptions for the period of extended operation.

4.1.1 Summary of Technical Information in the Application

To identify the TLAAs, the applicant evaluated calculations for VEGP against the six criteria specified in 10 CFR 54.3. The applicant indicated that it has identified the calculations that met the six criteria by searching the current licensing basis (CLB). The CLB includes the updated final safety analysis report (UFSAR), engineering calculations, technical reports, engineering work requests, licensing correspondence, and applicable vendor reports. In LRA Table 4.1.2-1, "List of VEGP Time-Limited Aging Analyses," the applicant listed the applicable TLAAs:

- reactor vessel neutron embrittlement
- metal fatigue
- environmental qualification of equipment
- concrete containment tendon prestress
- penetration load cycles
- leak before break
- diesel fuel oil storage tank corrosion allowance
- steam generator tube, loss of material
- cold overpressure protection system (COPS)

In a letter dated March 20, 2008, the applicant amended the LRA with LRA Section 4.7.5, "Underclad Cracking of the Reactor Pressure Vessel," which provides the applicant's time-limited aging analysis for managing underclad cracking in those reactor pressure vessel (RPV) components that are fabricated from SA 508, Class 2 forgings whose internal cladding was welded using a high heat submerged arc weld process.

Pursuant to 10 CFR 54.21(c)(2), the applicant identified exemptions granted under 10 CFR 50.12 based on TLAAs as defined in 10 CFR 54.3. The applicant compiled a list of the exemptions from searches of docketed correspondence, the operating licenses, and the

UFSAR, then evaluated each exemption in effect to determine whether it involved a TLAA as defined in 10 CFR 54.3. The applicant listed the following TLAA exemptions in LRA Section 4.1.3, "Identification of Exemptions:"

This evaluation found one exemption meeting the TLAA definition and two not based on TLAAs but permitting changes in TLAA methods. The analyses for these exemptions are included in LRA Table 4.1.2-1.

The first involves an exemption from the 10 CFR 50 Appendix A, General Design Criterion 4 requirement to assume a break "equivalent ... to the double-ended rupture of the largest pipe in the reactor coolant system." A letter dated February 5, 1985, granted VEGP an exemption that eliminated any need (1) to postulate pipe breaks in the primary loop, (2) to install primary loop jet impingement shields and pipe whip restraints, or (3) to consider dynamic effects (*i.e.*, jet impingement loads and blowdown loads in the primary loop and attached piping). LRA Section 4.7.1 describes the leak-before-break analysis.

The second exemption is from 10 CFR Part 50 Appendix G requirements for determining pressure-temperature (P-T) limit curves. This exemption eliminates the flange requirement based on the approved WCAP-16142-P, Revision 1 methodology. LRA Section 4.2.5 describes the P-T analysis.

The third exemption is from the requirements of 10 CFR 50.60, "Acceptance Criteria for Fracture Prevention for Light-Water Nuclear Power Reactors for Normal Operation," permitting the use of ASME Code Case N-514, "Low-Temperature Overpressure Protection" in lieu of the safety margins required by 10 CFR Part 50 Appendix G. LRA Section 4.7.4 describes the COPS setpoint analyses.

4.1.2 Staff Evaluation

LRA Section 4.1 lists the VEGP TLAAs; the applicant also discussed exemptions based on these TLAAs. The staff reviewed the information to determine whether the applicant has provided sufficient information pursuant to 10 CFR 54.21(c)(1) and 10 CFR 54.21(c)(2).

As defined in 10 CFR 54.3, TLAAs meet the following six criteria:

(1) they involve systems, structures, and components within the scope of license renewal, as described in 10 CFR 54.4(a)

(2) they consider the effects of aging

(3) they involve time-limited assumptions defined by the current operating term (40 years)

(4) they are determined to be relevant by the applicant in making a safety determination

(5) they involve conclusions, or provide the basis for conclusions, related to the capability of the system, structure, and component to perform its intended functions, as described in 10 CFR 54.4(b)

(6) they are contained or incorporated by reference in the CLB

The applicant reviewed the list of common TLAAs in NUREG-1800, Revision 1, "Standard Review Plan for Review of License Renewal Applications for Nuclear Power Plants" (SRP-LR), dated September 2005. The applicant listed TLAAs applicable to VEGP in LRA Table 4.1.2-1.

As required by 10 CFR 54.21(c)(2), an applicant must list all exemptions granted under 10 CFR 50.12, based on a TLAA, and evaluated and justified for continuation through the period of extended operation. The LRA states that each active exemption was reviewed to determine whether the exemption was based on a TLAA. The applicant identified TLAA-based exemptions. Based on the information provided by the applicant regarding the process used to identify these exemptions and its results, the staff concludes that there is one TLAA-based exemption that is justified for continuation through the period of extended operation.

During the audit, the staff noted that polar cranes are within the scope of license renewal, as delineated as Table 2.2-1 of the LRA, but the LRA does not mention whether it requires a TLAA. Therefore, the staff asked the applicant to clarify whether the vendor for the VEGP polar cranes has performed a fatigue analysis for the cranes that is based on a recommendation for allowable number of maximum load lifts. The applicant stated that an evaluation has been performed and that it demonstrates that the actual load cycles for these cranes through the period of extended operation are well below the design limits. In addition, the applicant stated that the polar crane evaluation is based only on assumed loads and does not evaluate a time-dependent aging effect. The staff reviewed the applicant's response and noted that the evaluations for cranes do not meet the definition of time-limited aging analysis. On this basis, the staff finds the applicant's response acceptable.

4.1.3 Conclusion

On the basis of its review, the staff concludes that the applicant has provided an acceptable list of TLAAs, as required by 10 CFR 54.21(c)(1). The staff confirmed, as required by 10 CFR 54.21(c)(2), that the applicant has provided an evaluation, as discussed in LRA Section 4.7.1, that justifies the continuation of the exemption pursuant to 10 CFR 50.12 for the period of extended operation.

4.2 Reactor Vessel Neutron Embrittlement

The regulations that govern reactor vessel integrity are codified in 10 CFR Part 50:

- Section 50.60 of 10 CFR requires all light-water reactors to meet 10 CFR Part 50 Appendices G and H fracture toughness, pressure-temperature (P-T) limits, and material surveillance program requirements for the reactor coolant boundary.
- Section 50.61 of 10 CFR provides fracture toughness requirements for protection against pressurized thermal shock.

Neutron embrittlement describes changes in mechanical properties of reactor vessel (RV) materials in the vicinity of the reactor core beltline region, i.e.; defined by the upper and lower active core planes. The metric of neutron exposure is fluence, i.e., the time integral of neutron flux with energies E>1.0 MeV. The most pronounced material change, relevant to this case, is reduction in fracture toughness with increasing fluence. As fracture toughness decreases with cumulative fast neutron exposure, the material's resistance to crack propagation decreases. Fracture toughness of ferritic materials depends upon temperature. The reference temperature for nil-ductility transition, RT_{NDT}, is the transition temperature above which the material is ductile, and below which is brittle. As neutron fluence increases, the RT_{NDT} increases and higher temperatures are required for the material to remain ductile. This shift in reference temperature is denoted as adjusted reference temperature (ART_{NDT}) and equals the sum $RT_{NDT} + \Delta RT_{NDT}$ where ΔRT_{NDT} is the difference induced by the fluence exposure. Determination of the projected RV reduction in fracture toughness as a function of neutron fluence affects several analyses that support Vogtle Nuclear Plant, Units 1 and 2 (VEGP) operations:

- RV Adjusted Reference-Temperature
- RV Material Upper-Shelf Energy (USE)
- RV Pressurized Thermal Shock (PTS)
- Pressure-Temperature Limits (PT Limits)

As extension of the operating period from 40 years to 60 years will increase neutron fluence, the 60-year fluence value and its impact upon the analyses that support operation must be determined.

4.2.1 Neutron Fluence

4.2.1.1 Summary of Technical Information in the Application

The applicant states that the regulatory requirements in 10 CFR 50.60 and 50.61 are supplemented with 10 CFR 54.21(c)(1)(ii) regarding the required information for TLAA items. The methodology employed in the derivation of the fluence values used in 10 CFR 50.61, the pressure-temperature limits and the low temperature overpressure protection (LTOP) settings are to adhere to the guidance in RG 1.190.

4.2.1.2 Staff Evaluation

The regulatory requirements in 10 CFR 50.60 and 50.61 are supplemented with 10 CFR 54.21(c)(1)(ii) regarding the required information for TLAA items. The methodology employed in the derivation of the fluence values used in 10 CFR 50.61, the pressure-temperature limits and the LTOP limits settings should adhere to the guidance in Regulatory Guide (RG) 1.190. Finally, RG 1.99 for extrapolation of the fluence to ¼T and ¾T locations should be followed.

The staff reviewed the license renewal application (LRA) and found only a statement that the fluence calculation adhered to the guidance of RG 1.190. That was not deemed adequate to conclude adherence to the guidance with RG 1.190, or whether the values listed accounted for the previously approved power uprate, and the applicable number of effective full power years (EFPYs) of the calculated fluence values. To complete the required information for the review, during the audit, the staff requested: (1) a reference for the calculations (2) clarification for the listed fluence values as to whether they include the power uprate and (3) clarification of the applicable EFPYs for the listed fluence values.

The applicant provided WCAP-16736-P Revision 1, May 2007. In RAI 4.2.1-04 the staff stated that Table 4.2.2-1 of the LRA shows a surface fluence of 3.2×10^{19} n/cm^2 (E > 1.0 MeV) for VEGP Unit 1 and 3.02×10^{19} n/cm^2 (E > 1.0 MeV) for VEGP Unit 2 whereas, Table 6.1.2-5 on page 6-12 of WCAP-16736 shows a surface fluence of 3.24×10^{19} n/cm^2 (E > 1.0 MeV) for VEGP Unit 1 and Table 6.1.2-7 on page 6-14 of WCAP-16736 shows a surface fluence of 3.06×10^{19} n/cm^2 (E > 1.0 MeV) for VEGP Unit 2. The staff asked the applicant to clarify inconsistencies between this reference and fluence values listed in Table 4.2.1-1 of the submittal.

The applicant's response, provided by letter dated March 31, 2008, reaffirmed a commitment to update the Tables in section 4.2.1 of the LRA to reflect the 57 EFPY fluence values contained in WCAP-16376-P in the fluence values. This commitment addressed the NRC staff's concern regarding the noted inconsistencies, and the NRC staff examined the license renewal update dated June 26, 2008, and confirmed that the inconsistency has been eliminated.

The NRC staff also requested, in RAI 4.2.1-04, that the licensee clarify whether the fluences in Table 4.2.1-1 of the LRA which show surface fluences of 3.20×10^{19} n/cm^2 n/cm^2 (E > 1.0 MeV) for VEGP Unit 1 and 3.02×10^{19} n/cm^2 (E > 1.0 MeV) for VEGP Unit 2 include the 1.7% power uprate.

In the March 31, 2008, letter, the applicant stated that the values in Table 4.2.2-1 were calculated for the pre-MUR power level when the EOL would have been at 57 EFPYs. After the MUR implementation the corresponding EFPYs are 56.3. Therefore, the section 4.2.1 Tables are for 56.3 EFPYs while the WCAP-16736 values were calculated for 57 EFPYs. This is acceptable because the 56.3 EFPYs are still conservative with respect to the expected value of EFPYs. The licensee stated that the Section 4.2.1 Tables will be updated to reflect the MUR update fluence values.

The applicant further stated that section 4.2-1 of the LRA discusses the fluence calculations based on the methodology in WCAP-14040-A Revision 4 that has been approved by the staff and adheres to the guidance in RG 1.190. The methodology has been benchmarked extensively to external measured values, such as the PCA (Pool Critical Assembly at ORNL) and to Vogtle 1 and 2 surveillance capsule results from the eight capsules removed and measured from both units.

The projected values (required by 54.21(c)(1)(ii)) are conservatively estimated to 57 EFPYs of operation at the current power level or 56.3 EFPYs at the MUR power level. Because the method adheres to guidance in RG 1.190, and the projected 57 EFPYs (and 56.3 EFPYs) are conservative the proposed fluence values are acceptable.

The fluence values were calculated at the peak inside surface beltline azimuthal locations for both units (45° and 30° azimuthal for Units 1 and 2 respectively) and the corresponding locations at the extended beltlines. All values were extended to ¼T and ¾T locations using the attenuation function in RG 1.99, Section 1.1, and Revision 2. Because the method adheres to RG 1.99 it is acceptable. Therefore, the staff concludes that the fluence values listed in Tables 4.2.1-1 and 4.2.1-2 of the submittal is acceptable.

4.2.1.3 UFSAR Supplement

The applicant provided a UFSAR supplement summary description of its TLAA evaluation of neutron fluence analysis in LRA Section A.3.1.1. On the basis of its review of the UFSAR supplement, the staff concludes that the summary description of the applicant's actions to address neutron fluence limits is adequate.

4.2.1.4 Conclusion

Section 4.2.1 of the submittal discusses the fluence calculation that was based on the methodology in WCAP-14040-A Revision 4, which has been approved by the staff and adheres to the guidance in RG 1.190. The methodology has been benchmarked extensively to external measured values, such as the PCA (Pool Critical Assembly at ORNL) and to Vogtle 1 and 2 surveillance capsule results from the eight capsules removed and measured from both units. The projected values (as required by 54.21(c)(1)(ii)) are conservatively estimated to 57 effective full power years (EFPYs) of operation at the current power level or 56.3 EFPYs when the planned measurement uncertainty recapture (MUR) becomes effective. Because the method adheres to guidance in RG 1.190, the projected EFPYs are conservative, and the method has been benchmarked as required it is acceptable.

Fluence values were calculated at the peak inside surface beltline azimuthal locations for both units (45° and 30° for Units 1 and 2 respectively) and the corresponding locations at the extended beltlines. All values were extended to ¼T and ¾T locations using the attenuation function in RG 1.99, Section 1.1, and Revision 2. Because the method adheres to RG 1.99 it is acceptable. Therefore, the staff concludes that the fluence values listed in Tables 4.2.1-1 and 4.2.1-2 of the submittal are acceptable.

On the basis of its review, as discussed above, the staff concludes that the fluence values listed in Tables 4.2.1-1 and 4.2.1-2 of the applicant's submittal is acceptable. The staff also determined that the UFSAR supplement contains an appropriate summary description of the TLAA evaluation, as required by 10 CFR 54.21(d).

4.2.2 Upper Shelf Energy Analysis

4.2.2.1 Summary of Technical Information in the Application

LRA Section 4.2.2 summarizes the evaluation of the upper shelf energy (USE) analysis for the period of extended operation. The applicant states that fracture toughness is a measure of a material's resistance to crack propagation. Charpy V-notch tests indirectly estimate fracture toughness, and Charpy V-notch test results are measured in ft.-lbs. of absorbed energy. The more ductile a material, the higher the fracture toughness and the more ft.-lbs. of energy will be absorbed during the Charpy V-notch test. The fracture toughness of RV steels is temperature-dependent. At low temperatures, the vessel material toughness is relatively low and constant and the material behaves in a brittle fashion. Rising temperatures reach a point where the toughness increases rapidly until another plateau where the toughness is relatively high and constant. In this high toughness region, the material is ductile. These regions of the curve are the lower shelf, transition zone, and upper shelf, respectively. The applicant notes that Title 10 of the Code of Federal Regulations (CFR) Part 50, Appendix G contains screening criteria that limit the degree that the USE value for a RV material may be allowed to drop due to neutron radiation exposure. The regulation requires the initial RV material USE to be equal to or above 75 ft.-lb. and for the USE to be equal to or above 50 ft.-lb. throughout the licensed life of the vessel, unless lower values of USE can be demonstrated to provide margins of safety against fracture equivalent to those required by the Appendix G of the American Society of Mechanical Engineers (ASME) Code, Section XI.

The applicant also states that an analysis of the USE of the VEGP's RV beltline materials for the license renewal period [56.3 effective full power years (EFPY)] requires the use of Regulatory Guide (RG) 1.99, Revision 2, "Radiation Embrittlement of Reactor Vessel Materials." The applicant further states that RV USE analyses were at the ¼T wall location of each beltline material using the respective copper contents and Figure 2 of the RG 1.99, Revision 2. The RV beltline material with the lowest predicted USE (61 ft.-lbs.) at 56.3 EFPY for VEGP, Unit 1 is Nozzle-to-Intermediate Shell Circumferential Weld 103-121 and the RV material with the lowest predicted USE (56 ft.-lbs.) at 56.3 EFPY for VEGP, Unit 2 is Shell Course Weld 105-121A. The projected USE values of these materials for the period of extended operation will remain above the required 50 ft-lb limit. This analysis demonstrates that, for the most limiting material in each RV, the lowest predicted USE is greater than the 10 CFR Part 50, Appendix G limit of 50 ft-lbs. The applicant states that, "Since all base materials and welds exceed the acceptance criteria of 10 CFR Part 50, Appendix G, these TLAA's have been shown to be acceptable in accordance with 10 CFR 54.21(c)(1)(ii)."

4.2.2.2 Staff Evaluation

The staff reviewed LRA Section 4.2.2 to verify pursuant to 10 CFR 54.21(c)(1)(ii) that the analysis has been projected to the end of the period of extended operation.

Section IV.A.1 to 10 CFR Part 50, Appendix G provides the Commission's requirements for demonstrating that RVs in U.S. light-water reactor facilities will have adequate protection from brittle failure throughout their service lives. The rule requires RV beltline materials to have USE values equal to or above 75 ft-lb when the materials are in the unirradiated condition and equal to or above 50 ft-lbs throughout the licensed life of the RV. RG 1.99, Revision 2 provides an expanded discussion regarding the calculations of USE values and describes two methods for determining USE values for RV beltline materials, depending on whether or not a given RV beltline material is represented in the plant's Reactor Vessel Material Surveillance Program.

The applicant provided its USE analyses for the RV beltline materials of VEGP, Units 1 and 2 in Tables 4.2.2-1 and 4.2.2-2, respectively, of the LRA. The USE analyses were based on the ¼T neutron fluence values listed in LRA Tables 4.2.1-1 and 4.2.1-1 and these neutron fluence values were based on the projected values at the end of the period of extended operation (i.e., at 56.3 EFPY). According to Table IV A-2 of NUREG-1801, Revision 1, "Generic Aging Lessons Learned Report," ferritic materials are subject to neutron embrittlement when they are exposed to a neutron fluence greater than 1×10^{17} n/cm^2 (E > 1 MeV) at the end of the period of extended operation.

The staff performed independent calculations of the USE values for the RV beltline materials through the expiration of the period of extended operation for VEGP, Units 1 and 2. The staff applied the ¼T neutron fluence values listed in LRA Table 4.2.1-1 for the VEGP, Unit 1 and 2 RVs as the basis for its independent calculations. The staff applied the methods provided in RG 1.99, Revision 2 for performing the independent USE calculations. The staff concludes that for VEGP, Unit 1 Nozzle-to-Intermediate Shell Plate Circumferential weld 103-21 is the limiting material and for VEGP, Unit 2 Shell Course Weld 105-121A is the limiting material. The staff calculated a USE value of 61 ft-lbs for the VEGP, Unit 1 weld at 0.0532×10^{19} n/cm^2 (E > 1.0 MeV) and this value is in good agreement with the 56.3 EFPY USE value (i.e., 61 ft-lbs) calculated by the applicant for this weld. The staff calculated a USE value of 56 ft-lb for the VEGP, Unit 2 weld at 0.0455×10^{19} n/cm^2 (E > 1.0 MeV) and this value is in good agreement with the 56.3 EFPY USE value (i.e., 56 ft-lbs) calculated by the applicant for this weld. Both of these values meet the acceptance criterion in 10 CFR Part 50, Appendix G for maintaining the USE values of the RV beltline materials above 50 ft-lbs throughout the licensed life of the plant. Therefore, since the bounding materials for the VEGP, Unit 1 and 2 RVs meet the requirements of 10 CFR Part 50, Appendix G, all of the VEGP, Unit 1 and 2 RV beltline materials meet the regulatory requirements.

Based on the technical assessments stated above, the staff concludes that the RVs at VEGP, Units 1 and 2 will maintain an acceptable level of USE values throughout the period of extended operation. The staff concludes that the applicant's TLAA for USE, as described in Section 4.2.2

of the LRA is in compliance with requirements of 10 CFR Part 50, Appendix G and, therefore, is acceptable.

4.2.2.3 UFSAR Supplement

The applicant provided a UFSAR supplement summary description of its TLAA evaluation of upper shelf energy analysis in LRA Section A.3.1.2. On the basis of its review of the UFSAR supplement, the staff concludes that the summary description of the applicant's actions to address upper shelf energy analysis is adequate.

4.2.2.4 Conclusion

On the basis of its review, as discussed above, the staff concludes that the applicant has demonstrated, pursuant to 10 CFR 54.21(c)(1)(ii), that the USE analysis for VEGP, Units 1 and 2 has been projected to the end of the period of extended operation. The staff also determined that the UFSAR supplement contains an appropriate summary description of the TLAA evaluation, as required by 10 CFR 54.21(d).

4.2.3 Pressurized Thermal Shock

4.2.3.1 Summary of Technical Information in the Application

LRA Section 4.2.3 summarizes the evaluation of the unit's pressurized thermal shock (PTS) analysis for the period of extended operation. The applicant states that 10 CFR 50.61 defines screening criteria for the embrittlement of RV materials in pressurized water reactors (PWRs) as well as actions required if these screening criteria are exceeded. The RV reference temperature for PTS, RT_{PTS}, will increase due to increasing neutron fluence, and the screening criteria specify limits on the RT_{PTS} values. The rule requires the RT_{PTS} values for all beltline materials to be maintained below the PTS screening criteria throughout the period of extended operation. For circumferential welds, the PTS screening criterion is 300°F. For plates, forgings, and axial welds the PTS screening criterion is 270°F.

The applicant also states that it performed a PTS evaluation for the RV beltline materials in accordance with 10 CFR 50.61. Calculation of RT_{PTS} values is by addition of the initial RT_{NDT}, the predicted radiation-induced change in material properties (ΔRT_{NDT}), and a margin term (m) to account for uncertainties in the values of initial RT_{NDT}, copper and nickel contents, neutron fluence, and calculation procedures. Calculation of the predicted radiation-induced ΔRT_{NDT} is by use of the respective RV beltline material copper and nickel contents and the neutron fluence applicable to the RV material through 56.3 EFPY of operation.

Evaluations of the RT_{PTS} values for each RV beltline material were based on the tabulated chemistry factor values described in 10 CFR 50.61.

The applicant also states that RT_{PTS} values for the RV beltline materials at 56.3 EFPY were determined and the results of the PTS evaluation demonstrated that the RV beltline materials will not exceed the PTS screening criteria before the end of the period of extended operation. The controlling RV beltline material for VEGP, Unit 1 is Intermediate Shell Plate Heat Number B8805-2, with an RT_{PTS} value of 123°F at 56.3 EFPY, which is well below the PTS screening criterion of 270°F. The controlling RV beltline material for VEGP, Unit 2 is Nozzle Shell Course R3-3, with an RT_{PTS} value of 134°F at 56.3 EFPY, which is also well below the PTS screening criterion of 270°F. The applicant states that, "Since all base materials and welds meet the screening criteria contained in 10 CFR Part 50.61 at EOL, these TLAA's have been shown to be acceptable for the period of extended operation in accordance with 10 CFR 54.21(c)(1)(ii)."

4.2.3.2 Staff Evaluation

The staff reviewed LRA Section 4.2.3 to verify, pursuant to 10 CFR 54.21(c)(1)(ii), that the analysis has been projected to the end of the period of extended operation.

10 CFR 50.61 provides the Commission's requirements for demonstrating that RVs in U.S. PWR facilities will have adequate protection against the consequences of PTS events throughout their licensed operating period. The rule requires licensees to calculate RT_{PTS} values for each base metal and weld material located in the beltline region of the RVs. The rule sets a screening limit of 270°F for RT_{PTS} values that are calculated for base metals (i.e., forging and plate materials) and axial weld materials and a screening limit of 300°F for RT_{PTS} values that are calculated for circumferential weld materials. The rule also provides an expanded discussion regarding how the calculations of RT_{PTS} values should be performed and describes two methods for determining RT_{PTS} values for RV beltline materials, depending on whether or not a given RV beltline material is represented in the plant's Reactor Vessel Material Surveillance Program.

The applicant provided its RT_{PTS} value assessments for the VEGP, Unit 1 and 2 RV beltline materials in Table 4.2.3-1 for VEGP, Unit 1 and Table 4.2.3-2 for VEGP, Unit 2 of the LRA. The RT_{PTS} values listed in these tables were based on the neutron fluence values at the clad-to-metal interface of the RV. According to Table IV A-2 of NUREG-1801, Revision 1, ferritic materials are subject to neutron embrittlement when they are exposed to a neutron fluence greater than 1×10^{17} n/cm^2 (E > 1 MeV) at the end of the period of extended operation. The applicant's neutron fluence values used to determine the RT_{PTS} values were based on the values that were projected to end of the period of extended operation (i.e., at 56.3 EFPY). The applicant reported that for VEGP, Unit 1 Intermediate Shell Plate Heat Number B8805-2 is the limiting material for PTS with a RT_{PTS} value of 123°F at 56.3 EFPY. The applicant reported that for VEGP, Unit 2 Nozzle Shell Course R3-3 is the limiting material for PTS with a RT_{PTS} value of 134°F at 56.3 EFPY. These values were calculated using the chemistry factors obtained from the chemical composition of the limiting beltline material and the tables in 10 CFR 50.61.

To verify the validity of the applicant's calculation of the RT_{PTS} values at 56.3 EFPY for VEGP's limiting beltline materials, the staff performed independent calculations per 10 CFR 50.61 and found the RT_{PTS} values acceptable. The staff confirmed that Intermediate Shell Plate Heat Number B8805-2 and Nozzle Shell Course R3-3 were the limiting beltline materials for VEGP Units 1 and 2, respectively. The staff calculated an RT_{PTS} value of 123.3°F for VEGP, Unit 1 Intermediate Shell Plate Heat Number B8805-2 and an RT_{PTS} value of 134.2°F for VEGP, Unit 2 Nozzle Shell Course R3-3. The staff finds the RT_{PTS} values for all VEGP, Unit 1 and 2 RV beltline materials to be acceptable because the bounding materials comply with the requirements specified in 10 CFR 50.61.

Based on the technical assessments stated above, the staff concludes that the RV's at VEGP will maintain acceptable RT_{PTS} values throughout the period of extended operation. The staff therefore determined that the applicant's TLAA for PTS, as described in Section 4.2.3 of the LRA, is in compliance with the screening criteria specified in 10 CFR 50.61. Therefore, the staff concludes that the VEGP, Unit 1 and 2 RVs will be acceptable for PTS throughout the period of extended operation.

4.2.3.3 UFSAR Supplement

The applicant provided a UFSAR supplement summary description of its TLAA evaluation of pressurized thermal shock analysis in LRA Section A.3.1.3. On the basis of its review of the UFSAR supplement, the staff concludes that the summary description of the applicant's actions to address pressurized thermal shock analysis is adequate.

4.2.3.4 Conclusion

On the basis of its review, as discussed above, the staff concludes that the applicant has demonstrated, pursuant to 10 CFR 54.21(c)(1)(ii), that the VEGP Units 1 and 2 RV PTS analysis has been projected to the end of the period of extended operation. The staff also determined that the UFSAR supplement contains an appropriate summary description of the TLAA evaluation, as required by 10 CFR 54.21(d).

4.2.4 Adjusted Reference Temperature

4.2.4.1 Summary of Technical Information in the Application

LRA Section 4.2.4 summarizes the evaluation of material adjusted reference temperature (ART) values for the period of extended operation. The applicant states that ART is the value of a material's Initial RT_{NDT} plus ΔRT_{NDT} plus margins for uncertainties at a specific location. Neutron embrittlement increases a material's ART value; thus, the minimum temperature at which an RV is allowed to be pressurized increases over the licensed period. The ART value of the limiting beltline material is used to correct the RV beltline P-T limits to account for radiation effects.

The applicant also states that calculation of ART values for the RV beltline region materials in accordance with RG 1.99, Revision 2 was performed by the addition of the initial RT_{NDT} to the predicted radiation-induced ΔRT_{NDT} and a margin term to account for uncertainties in the values of initial RT_{NDT}, copper and nickel contents, fluence, and the calculation procedures. Calculation of the predicted radiation-induced ΔRT_{NDT} was done by using the respective RV beltline material copper and nickel contents for, and the neutron fluence applicable to, a particular RV beltline material at 56.3 EFPY. The applicant further states that the evaluation for the ART values were performed at the ¼T and ¾T wall locations of each RV beltline material with chemistry factors determined from the tabulated values described in RG 1.99, Revision 2. In this manner, ART values for the RV beltline region materials applicable to 56.3 EFPY were determined. These results show that the RV limiting beltline materials at both ¼T and ¾T locations are Intermediate Shell Plate Heat Number B8805-2 for VEGP, Unit 1 and Lower Shell Plate R8-1 for VEGP, Unit 2. The applicant updated these calculations for EOL through the period of extended operation in accordance with 10 CFR 54.21(c)(1)(ii).

4.2.4.2 Staff Evaluation

The staff reviewed LRA Section 4.2.4 to verify, pursuant to 10 CFR 54.21(c)(1)(ii), that the analysis of ART values has been projected to the end of the period of extended operation. The staff reviewed the ART values listed in Tables 4.2.3-1 and 4.2.3-2 of the LRA and performed independent calculations of the ART values for the RV beltline materials by using the method specified in RG 1.99, Revision 2. The ART values of the limiting beltline materials, Intermediate Shell Plate Heat Number B8805-2 for VEGP, Unit 1 and Lower Shell Plate R8-1 for VEGP, Unit 2, were verified per RG 1.99, Revision 2. The staff independently calculated ART values of 116.3°F and 101.3°F at the ¼T and ¾T locations for VEGP, Unit 1 Intermediate Shell Plate

Heat Number B8805-2. The staff independently calculated ART values of 126.6°F and 112.5°F at the ¼T and ¾T locations for VEGP, Unit 2 Lower Shell Plate R8-1. The staff's calculations show that these materials were bounding for all VEGP, Units 1 and 2 RV beltline materials. The staff finds the ART values for these materials to be acceptable as they comply with the requirements of 10 CFR Part 50, Appendix G and the guidance of RG 1.99, Revision 2.

Based on the technical assessments stated above, the staff concludes that the ART values for the RV beltline materials, as projected through the period of extended operation are in conformance with the recommended guidelines of RG 1.99, Revision 2. Therefore, the staff concludes that the applicant's TLAA for the VEGP ART values is acceptable.

4.2.4.3 UFSAR Supplement

The applicant provided a UFSAR supplement summary description of its TLAA evaluation of adjusted reference temperature in LRA Section A.3.1.4. On the basis of its review of the UFSAR supplement, the staff concludes that the summary description of the applicant's actions to address adjusted reference temperature analysis is adequate.

4.2.4.4 Conclusion

On the basis of its review, as discussed above, the staff concludes that the applicant has demonstrated, pursuant to 10 CFR 54.21(c)(1)(ii), that the analysis of RV beltline ART values has been projected to the end of the period of extended operation. The staff also determined that the UFSAR supplement contains an appropriate summary description of the TLAA evaluation, as required by 10 CFR 54.21(d).

4.2.5 Pressure-Temperature Limits

4.2.5.1 Summary of Technical Information in the Application

LRA Section 4.2.5 summarizes the evaluation of operating pressure-temperature (P-T) limits for the period of extended operation. The applicant states that neutron embrittlement increases the ART; thus, the minimum temperature at which an RV is allowed to be pressurized increases over the licensed period. The ART value for the limiting beltline material is used to correct the beltline P-T limits to account for radiation effects. In accordance with 10 CFR Part 50, Appendix G, RV thermal limit analyses must determine operating P-T limits for boltup, hydrotest, pressure tests, normal operation, and anticipated operational occurrences. P-T operating limits are specifically required for three categories of operation: (1) hydrostatic pressure tests and leak tests, (2) non-nuclear heat-up/cool-down and low-level physics tests, and (3) core critical operation.

RV P-T limits and minimum temperature requirements in accordance with 10 CFR Part 50, Appendix G are defined by operating condition, vessel pressure, presence of fuel in the vessel, and core criticality. The P-T limits must be at least as conservative as limits obtained by the methods of analysis and margins of safety of Appendix G of the ASME Code, Section XI. The minimum temperature requirements pertain to the limiting material, which is either the highly stressed material in the closure flange region or a material in the beltline region with the highest ART value.

The applicant also states that VEGP, Units 1 and 2 are currently operating to 36 EFPY P-T limit curves, which are included in the Pressure and Temperature Limits Report (PLTR) for each unit. As described in the PTLR, the Reactor Vessel Surveillance Program updates the P-T limit curves considering the data gained from capsules SNC pulls, and the content and update of the PTLR is in accordance with the VEGP Technical Specification 5.6.6. When the operating conditions of each unit merit the use of a different curve, the PTLR for that unit will be updated to include P-T limit curves that bound the current level of neutron embrittlement (i.e., EFPY) for the unit. Therefore, the applicant further states that this TLAA is in accordance with 10 CFR 54.21(c)(1)(ii) and 10 CFR 54.21(c)(1)(iii).

4.2.5.2 Staff Evaluation

The staff reviewed LRA Section 4.2.5, to verify pursuant to 10 CFR 54.21(c)(1)(ii), that the analysis has been projected to the end of the period of extended operation.

Paragraph IV.A.2 of 10 CFR Part 50, Appendix G provides the staff's requirements and criteria for generating the P-T limits that are required for commercial U.S. light-water reactors. 10 CFR 50.36 requires licensees owning nuclear power production facilities to include the P-T limits and low pressure over-pressure protection (LTOP) system set points among the limiting conditions for operation (LCOs) in plant Technical Specifications (TS).

The staff, in a letter dated March 28, 2005, approved the current VEGP P-T limits which are valid for 36 EFPY. Revision of the P-T limits is based on the extent to which the beltline materials are exposed to the neutron fluence during the period of extended operation. Regarding the applicant's use of surveillance capsule test data, the LRA states, "As described in the PTLR, the Reactor Vessel Surveillance Program (Appendix B.3.25) updates the P-T limit curves considering the data gained from capsules SNC pulls, and the content and update of the PTLR is in accordance with the VEGP Technical Specification 5.6.6. When the operating conditions of each unit merit the use of a different curve, the PTLR for that unit is updated to include P-T limit curves that bound the current level of neutron embrittlement (i.e., EFPY) for the unit."

The staff finds the applicant's statement to manage the P-T limits acceptable because the change in P-T limits will be implemented in accordance with the facility's current licensing basis which meets the regulatory requirements of 10 CFR 50.60 and 10 CFR Part 50, Appendix G.

Based on the technical assessments stated above, the staff concludes that the applicant's TLAA for the pressure-temperature (P-T) limits are acceptable.

4.2.5.3 UFSAR Supplement

The applicant provided a UFSAR supplement summary description of its TLAA evaluation of the pressure temperature limits analysis in LRA Section A.3.1.5. On the basis of its review of the UFSAR supplement, the staff concludes that the summary description of the applicant's actions to address pressure temperature limits analysis is adequate.

4.2.5.4 Conclusion

On the basis of its review, as discussed above, the staff concludes that the applicant has demonstrated, pursuant to 10 CFR 54.21(c)(1)(ii) and 10 CFR 54.21(c)(1)(iii), that the applicant's processes will adequately manage the VEGP, Units 1 and 2 P-T limit curves for the period of extended operation. In addition, the applicant has demonstrated pursuant to 10 CFR 54.21(c)(1)(ii) that the TLAA has been projected to the end of the period of extended operation.

The staff also determined that the UFSAR supplement contains an appropriate summary description of the TLAA evaluation, as required by 10 CFR 54.21(d).

4.3 Metal Fatigue

The VEGP design basis addresses the effects of metal fatigue. Fatigue is an age-related degradation mechanism caused by either mechanical or thermal cyclic stressing of a component. Fatigue analyses are TLAAs if they meet all six elements of the 10 CFR 54.3(a) definition. If the analyses are based on a number of cycles estimated for the current license term, they may meet criterion 54.3(a)(3) as based on the current operating term. If a component has a fatigue TLAA that remains valid for (demonstration in accordance with 10 CFR 54.21 (c)(1)(i)) or is projected to the end of (demonstration in accordance with 10 CFR 54.21 (c)(1)(ii)) the period of extended operation, cracking due to metal fatigue is not an aging effect requiring management for that component. If the fatigue TLAA cannot demonstrably remain valid for the period of extended operation by either of these methods, an aging management program (AMP) must manage the fatigue (demonstration in accordance with 10 CFR 54.21(c)(1)(iii)).

4.3.1 Fatigue of ASME Class 1 Components

The VEGP design incorporates the requirements of ASME Code Section III Class 1, which requires a discrete analysis of the thermal, mechanical, and dynamic stress cycles on reactor coolant pressure boundary components. The staff noted that although original design specifications commonly state that the transient conditions are for a 40-year design life, the fatigue analyses are based on the specified number of transient occurrences rather than on this lifetime. The applicant selected a design number for each transient somewhat larger than expected to occur during the 40-year licensed life of the plant based on operating experience and on projections of future operation based on system designs.

In addition to the original design transients, fatigue loading transients and issues subsequently identified are not parts of the original fatigue analyses. The staff noted that stratification and insurge-outsurge transients are evaluated for the lower pressurizer head and surge line, and the impact of the reactor coolant system (RCS) environment on the fatigue life of piping and components requires specific evaluation for license renewal.

4.3.1.1 Class 1 Piping and Component Design Transient Cycles

4.3.1.1.1 Summary of Technical Information in the Application

LRA Section 4.3.1.1 summarizes the evaluation of Class 1 piping and component design transient cycles for the period of extended operation. UFSAR Table 3.9.N.1-1 summarizes the RCS design transients. LRA Table 4.3.1-1 lists these transients and indicates those under the Fatigue Monitoring Program. The applicant states that this program monitors transients and components to assure that actual plant experience remains bounded by design analyses assumptions. This program counts cycles of design-basis transient events and evaluates the number of occurrences against the design basis. LRA Table 4.3.1-1 indicates the basis for transients not monitored.

LRA Table 4.3.1-2 lists the component cyclic or transient limits that require monitoring as listed in UFSAR Table 3.9.N.1-2, including the design-basis cycles for each transient monitored, the number of cycles experienced as of October 9, 2005, and projected cycles based on current cycles to date to demonstrate that these transients will not exceed the design cycles in 60 years.

For the feedwater cycling, loss of charging flow (loop 1, loop 4), and loss of letdown and return to service transients, the CLB relies on cumulative usage factor (CUF) monitoring using stress-based fatigue monitoring of the limiting component locations (as indicated in the table). These component locations are the steam generator (SG) main and auxiliary feedwater nozzles and the normal and alternate charging nozzles. LRA Sections 4.3.1.2 and 4.3.1.3 describe and LRA Table 4.3.1-2 includes the evaluation of the projected CUF for the period of extended operation for these locations.

The CLB also relies on stress-based fatigue monitoring for the pressurizer surge line and lower pressurizer head including thermal stratification issues. LRA Section 4.3.1.4 describes the evaluation of the projected CUF for the period of extended operation.

4.3.1.1.2 Staff Evaluation

The staff reviewed LRA Section 4.3.1.1, to verify pursuant to 10 CFR 54.21(c)(1)(i), that the analyses remain valid for the period of extended operation and, pursuant to 10 CFR 54.21(c)(1)(iii), that the effects of aging on the intended function(s) will be adequately managed for the period of extended operation.

The staff reviewed Table 4.3.1-2, Transients Tracked by Fatigue Monitoring Software, and noted that inadvertent safety injection transient has zero projected cycles. The staff concludes that having a zero projected cycle is not conservative and asked the applicant, during the audit and review, to confirm that the inadvertent safety injection transient is not used in any baseline fatigue. In its response, the applicant stated that SNC manually changed projections that software calculated as zero to one, but missed this particular cycle. The applicant also stated the projected number of inadvertent safety injections is not used in any baseline fatigue analysis. In a letter dated March 20, 2008, the applicant amended the LRA by changing the projected cycle for inadvertent safety injections to one. The staff noted the transient's projected cycle does not affect the fatigue analysis. On this basis, the staff finds the applicant's response acceptable.

4.3.1.1.3 UFSAR Supplement

The applicant provided a UFSAR supplement summary description of its TLAA evaluation of metal fatigue in LRA Section A.3.2. On the basis of its review of the UFSAR supplement, the staff concludes that the summary description of the applicant's actions to address metal fatigue analysis is adequate.

4.3.1.1.4 Conclusion

On the basis of its review, as discussed above, the staff concludes that the applicant has demonstrated, pursuant to 10 CFR 54.21(c)(1)(i), that, for Class 1 piping and component design fatigue analysis, the analyses remain valid for the period of extended operation. The staff also determined that the UFSAR supplement contains an appropriate summary description of the TLAA evaluation, as required by 10 CFR 54.21(d).

4.3.1.2 CUF Monitoring - SG Main and Auxiliary Feedwater Nozzles

4.3.1.2.1 Summary of Technical Information in the Application

LRA Section 4.3.1.2 summarizes the evaluation of CUF monitoring - SG main and auxiliary feedwater nozzles for the period of extended operation. Analysis of the SGs was in accordance with ASME Code Section III Class 1 requirements. The applicant states that in this analysis, one of the transients in the SG fatigue analysis was 2000 cycles of intermittent (slug) feeding of 32 °F feedwater into the SG at hot standby or no-load conditions. These 2000 cycles also include feedwater additions required during plant heatup and cooldown operations. Based on current operating experience, the analysis projects several SGs to exceed this number of cycles.

The applicant also states that rather than simple cycle counting of feedwater cycling events, VEGP uses stress-based fatigue monitoring of the SG main and auxiliary feedwater nozzles, the SG locations affected most by these events. The applicant determined from the fatigue analysis for these nozzles that the CUF with these increased cycles would remain well below 1.0. LRA Table 4.3.1-2 shows the calculated and 60-year projected CUFs for the main and auxiliary feedwater nozzles. The applicant further states that the Fatigue Monitoring Program will manage these nozzles by stress-based fatigue monitoring so the CUF remains below 1.0.

4.3.1.2.2 Staff Evaluation

The staff reviewed LRA Section 4.3.1.2, to verify pursuant to 10 CFR 54.21(c)(1)(iii), that the effects of aging on the intended function(s) will be adequately managed for the period of extended operation.

The staff noted that LRA Table 4.3.1-2, Transients Tracked by Fatigue Monitoring Software, did not indicate the start date for the transients tracked. During the audit and review, in an effort to clarify the transient cycles counted, the staff asked the applicant to clarify the start date of cycling counting. In its response, the applicant provided a table that contains the transient events documented in Table 4.3.1-2 along with the date of first recorded events for both units. The staff reviewed the applicant's response as well as the Westinghouse Report SE-ICAT(96)-212, which contained a detailed description of available plant records and the baseline count established at the time for each cycle counted. From its review, the staff noted software data for transient cycles has been generated since June 30, 1995, and that the cycles accrued prior to that date were properly documented. On this basis, the staff finds the applicant's response acceptable. The staff finds that the Fatigue Monitoring Program will adequately manage the fatigue usage of the steam generator main and auxiliary nozzles during the period of extended operation.

4.3.1.2.3 UFSAR Supplement

The applicant provided a UFSAR supplement summary description of its TLAA evaluation of CUF monitoring - SG main and auxiliary feedwater nozzles in LRA Section A.3.2.1.

On the basis of its review of the UFSAR supplement, the staff concludes that the summary description of the applicant's actions to address CUF monitoring - SG main and auxiliary feedwater nozzles is adequate.

4.3.1.2.4 Conclusion

On the basis of its review, as discussed above, the staff concludes that the applicant has demonstrated, pursuant to 10 CFR 54.21(c)(1)(iii), that, for CUF monitoring - SG main and auxiliary feedwater nozzles, the effects of aging on the intended function(s) will be adequately managed for the period of extended operation. The staff also determined that the UFSAR supplement contains an appropriate summary description of the TLAA evaluation, as required by 10 CFR 54.21(d).

4.3.1.3 CUF Monitoring - Charging Nozzles

4.3.1.3.1 Summary of Technical Information in the Application

LRA Section 4.3.1.3 summarizes the evaluation of CUF monitoring - charging nozzles for the period of extended operation. The applicant analyzed Class 1 portions of the charging and letdown systems for 20 cycles each of charging trip with prompt return and of letdown trip with delayed return. The applicant states that the projected cycles of these two transients are greater than design; however, the magnitudes of the transients and their resultant fatigue contribution are smaller than those in the design analyses.

The applicant also states that VEGP uses stress-based fatigue monitoring of the normal and alternate charging nozzles, the bounding locations in the Class 1 portion of the charging and letdown systems, rather than simple cycle counting of loss of charging, loss of letdown events, or both to maintain design limits for components affected by these cycles.

The applicant further states that it determined from fatigue analysis for these nozzles that the CUF with these increased cycles would remain well below 1.0. LRA Table 4.3.1-2 shows the calculated and 60-year projected CUFs for the charging nozzles, which the Fatigue Monitoring Program will manage by stress-based fatigue monitoring so the CUF remains below 1.0, the lowest established limit (LRA Section 4.3.1.7).

4.3.1.3.2 Staff Evaluation

The staff reviewed LRA Section 4.3.1.3, to verify pursuant to 10 CFR 54.21(c)(1)(iii), that the effects of aging on the intended function(s) will be adequately managed for the period of extended operation.

The LRA indicates that projected cycles of letdown trip with delayed return and charging trip with prompt return are greater than design. The LRA further stated that the magnitude of the transients and resultant fatigue contribution is smaller than in the design analysis. During the audit and review, the staff asked the applicant to explain how the magnitude of the transients and resultant fatigue is smaller than the design analysis and to provide all supporting documentation.

In its response, the applicant stated the temperature changes for actual events are less severe in magnitude and rate than the design transients. Furthermore, the applicant stated the increase in the number of cycles is small compared to the difference in severity between actual events and the design transients. The applicant concluded that since the actual temperature changes are less severe than the design transients and the increase in events is small, it follows that the actual CUF is less than the design CUF.

The staff reviewed the applicant's response as well as its supporting analysis reports, which define the severity of both transients and contain plots of actual transients. The staff noted the applicant uses stress-based monitoring, which include monitoring temperature changes instead of cycle counting. The staff confirmed that temperature changes are less in magnitude than that of design transients. Therefore, the staff agreed that applicant's statement is reasonable considering monitored transients data. On this basis, the staff finds the applicant's response acceptable.

4.3.1.3.3 UFSAR Supplement

The applicant provided a UFSAR supplement summary description of its TLAA evaluation of CUF monitoring - charging nozzles in LRA Section A.3.2.1. On the basis of its review of the UFSAR supplement, the staff concludes that the summary description of the applicant's actions to address CUF monitoring - charging nozzles is adequate.

4.3.1.3.4 Conclusion

On the basis of its review, as discussed above, the staff concludes that the applicant has demonstrated, pursuant to 10 CFR 54.21(c)(1)(iii), that, for CUF monitoring - charging nozzles, the effects of aging on the intended function(s) will be adequately managed for the period of extended operation. The staff also determined that the UFSAR supplement contains an appropriate summary description of the TLAA evaluation, as required by 10 CFR 54.21(d).

4.3.1.4 Thermal Stratification of the Surge Line and Lower Pressurizer Head

4.3.1.4.1 Summary of Technical Information in the Application

LRA Section 4.3.1.4 summarizes the evaluation of thermal stratification of the surge line and lower pressurizer head for the period of extended operation. The applicant evaluated the impact of thermal stratification on the fatigue usage in the surge line in support of its response to NRC Bulletin 88-11. That evaluation found the surge line fatigue usage acceptable for 40 years of operation with thermal stratification effects due to insurge and outsurge from the pressurizer. For license renewal, the applicant states that the staff noted that the stress-based fatigue monitoring software module for the surge line calculates the actual CUF due to changes in temperature, pressure, and other parameters of surge line and lower pressurizer head fatigue and accounts for pressurizer insurge-outsurge and thermal stratification effects in both the lower pressurizer head and both surge line nozzles. The applicant also states that the highest 60-year projected CUF for these components is 0.06 for the Unit 2 hot leg surge nozzle; thus, the applicant addresses Westinghouse Commercial Atomic Power (WCAP)-14574A Renewal Applicant Action Item 3.3.1.1.-1 for license renewal by using the stress-based monitoring software for the pressurizer lower head and surge line nozzles and demonstrating that design limits are maintained throughout the period of extended operation. The applicant further states that the Fatigue Monitoring Program will manage the pressurizer lower head and surge line nozzles by stress-based fatigue monitoring so the CUF remains below 1.0, the lowest established limit (LRA Sections 4.3.1.5.3 and 4.3.1.6).

4.3.1.4.2 Staff Evaluation

The staff reviewed LRA Section 4.3.1.4, to verify pursuant to 10 CFR 54.21(c)(1)(iii), that the effects of aging on the intended function(s) will be adequately managed for the period of extended operation.

The staff reviewed section LRA Section 4.3.1.4 and noted that the LRA states that pressurizer lower head and surge line nozzle will be managed using stress-based fatigue monitoring by the Fatigue Monitoring Program. However, the applicant did not address the pressurizer lower head. In RAI 4.3-1, the staff asked the applicant to provide the limiting 60-year projected CUF value for the pressurizer lower head. In a letter dated February 21, 2008, the applicant provided the limiting 60-year projected CUF value, 0.00017, for the pressurizer lower head.

The staff noted that this value is lower than the ASME code limit of 1.0. On this basis, the staff finds the applicant's response acceptable.

In addition, the staff reviewed an NRC safety evaluation entitled "Vogtle Unit 1 Safety Evaluation on Pressurizer Surge Line Thermal Stratification," dated April 12, 1990. In this SER, the staff noted that it states:

> "Applicant committed to revise applicable operating procedures to limit the system delta T (between the pressurizer head and the reactor coolant loop) for reactor coolant system (RCS) heatup (HU) to 320°F and RCS cool down (CD) to 300°F. The revised heatup and cooldown procedures ensure consistency between actual plant operation and the surge line analysis assumption."

In RAI 4.3-1, the staff also asked the applicant to discuss the procedures that have been used by VEGP. In addition, the staff asked the applicant to demonstrate the consistency between the recorded plant operational transient data and the assumptions that were made and used in the surge line and pressurizer lower head thermal stratification analyses.

In a letter dated February 21, 2008, the applicant provided the procedures numbers for cooldown to cold shutdown and heatup to hot shutdown. Both procedures, the applicant states, include a caution which calls for maintaining the Delta-T between the RCS and the pressurizer space as low as practical. In addition, both procedures state that the delta-T of 320 and 300 should not be exceeded for heatup and cool down, respectively. The applicant also provided the delta-T data from 1/1/97 through 2/18/08, which revealed that no delta-T for heatups and cooldowns reached 320 F and 300 F, respectively. In addition, the applicant states that in general, the maximum delta-T for each heatup and cooldown was in the 240 F and 260 F range. The staff reviewed applicant's response and noted that data available is within the limits, and therefore satisfies the commitment made by the applicant for the SER dated April 12, 1990. On this basis, the staff finds the applicant's response acceptable.

4.3.1.4.3 UFSAR Supplement

The applicant provided a UFSAR supplement summary description of its TLAA evaluation of thermal stratification of the surge line and lower pressurizer head in LRA Section A.3.2.1. On the basis of its review of the UFSAR supplement, the staff concludes that the summary description of the applicant's actions to address thermal stratification of the surge line and lower pressurizer head is adequate.

4.3.1.4.4 Conclusion

On the basis of its review, as discussed above, the staff concludes that the applicant has demonstrated, pursuant to 10 CFR 54.21(c)(1)(iii), that, for thermal stratification of the surge line and lower pressurizer head, the effects of aging on the intended function(s) will be adequately managed for the period of extended operation. The staff also determined that the UFSAR supplement contains an appropriate summary description of the TLAA evaluation, as required by 10 CFR 54.21(d).

4.3.1.5 *Effects of Reactor Coolant System Environment on Fatigue Life of Piping and Components*

4.3.1.5.1 Summary of Technical Information in the Application

LRA Section 4.3.1.5 summarizes the evaluation of effects of the RCS environment on fatigue life of piping and components for the period of extended operation. Generic Safety Issue 190 addressing fatigue life of metal components was closed in December 1999. In the closure letter, however, the NRC concluded that licensees should address the effects of reactor coolant environment on the fatigue life of selected components as they formulate AMPs in support of license renewal. The applicant stated that it evaluated these environmentally-assisted fatigue effects for plant-specific locations equivalent to those of NUREG/CR-6260 Section 5.4 for the newer vintage Westinghouse plant. These locations are reactor vessel shell and lower head, reactor vessel inlet and outlet nozzles, surge line hot leg nozzle, charging nozzle, and safety injection nozzle. The applicant also stated that it used formulas from NUREG/CR-5704 for stainless steel and from NUREG/CR-6583 for carbon and low-alloy steel components and showed the projected CUF values for these locations.

4.3.1.5.2 Staff Evaluation

The staff reviewed LRA Section 4.3.1.5, to verify pursuant to 10 CFR 54.21(c)(1)(i), that the analyses remain valid for the period of extended operation and, pursuant to 10 CFR 54.21(c)(1)(iii), that the effects of aging on the intended function(s) will be adequately managed for the period of extended operation.

LRA Section 4.3.1.5 states that cooldown/heatup cycles for Unit 1 and Unit 2 from June 30, 1995, through October 9, 2005, were analyzed to determine the average CUF. The staff noted this period is not the entire operating period of the plant, and asked the applicant to explain how this value was projected to the 60-year CUF during the audit and review.

In its response, the applicant stated that the transient projections are made using a weighting methodology that considers all cycles that have occurred but gives more weight to cycles that have occurred more recently than those that occurred in the past. The applicant provided the formula for calculating projections. In addition, the applicant stated the projected cycles of the transients and the resultant projected CUF are updated every 18 months so a future increase in frequency of a given transient will be quickly incorporated into a new projected number of cycles that gives more conservative values than a traditional straight line average projection.

The staff reviewed the applicant's response and reviewed a recent 18-month Vogtle fatigue monitoring report, which contain updates to VEGP cycle and CUF projections. The staff noted updating to the VEGP cycle and CUF projections with weighting methodology provide a realistic and current projection of transient-cycles. On this basis, the staff finds the applicant's response acceptable.

The staff reviewed the operating history of Vogtle and noted the plant implemented a modified operating procedure (MOP) to mitigate pressurizer insurge/outsurge transients. In RAI 4.3-2, the staff asked the applicant to discuss this MOP in detail and explain how the impacts of MOP were factored into the calculation of the average CUF per HU/CD in the EAF analysis.

In a letter dated February 21, 2008, the applicant provided its response to RAI 4.3-2, which includes the changes in plant procedures (heatup to hot shutdown, heatup to normal operating temperature and pressure, and cooldown to cold shutdown), findings of WCAP-14950, and adjustment the applicant will make to the CUF values for heater penetrations and surge nozzles.

The applicant stated that aspects of MOP were established at VEGP early in plant life. The applicant indicated that precautions were established in Rev. 0 of procedures (heatup to hot shutdown, heatup to normal operating temperature and pressure, and cooldown to cold shutdown) to limit the difference in boron concentration between the pressurizer and RCS, which results in limiting pressurizer insurges. In addition, the applicant indicated that another precaution was added to minimize the delta-T between the RCS hot leg and the pressurizer to address thermal stratification of the surge line.

The applicant provided overall strategies to mitigate insurge/outsurge transients in WCAP-14950, which includes 1) continuous pressurizer outsurge flow during heatup and cooldown operation and 2) minimizing the system delta-T. In addition, the applicant also provided CUF comparison between the standard steam bubble method and the modified steam bubble method documented in WCAP-14950. The WCAP report shows that for heater penetration and surge nozzle knuckle (bounding locations for pressurizer lower head and surge nozzle), the CUF was 1.84 and 1.88 times higher for the standard steam bubble method than the modified steam bubble method.

The applicant stated that VEGP now uses the modified steam bubble method, and that original procedures allowed either the standard or the modified steam bubble method. In addition, the applicant revised the 60-year CUF projection for heater penetration, surge nozzle, and hot leg surge nozzle by doubling the CUF values (that were calculated to account for transients that occurred from the first date of operation until 1/1/1998 for Unit 1 and 6/30/1995 for Unit 2) for these locations. In letters dated March 20, 2008 and June 27, 2008, the applicant amended its LRA by updating the CUF projection values for the locations mentioned.

The staff reviewed applicant's response and noted that applicant implemented the WCAP strategies early in plant life. Thus, the staff finds that the applicant minimized the number of insurge and outsurge transients before the issuance of WCAP-14950 and the subsequent MOP. In addition, the staff finds that by amending the LRA to account for higher CUF values due to standard steam bubble procedure that may be used prior to implementation of monitoring software, the new CUF values now accounted for the possible use of the standard bubble method. On these bases, the staff finds the applicant's response acceptable.

The staff reviewed LRA Sections 4.3.1.5.4 and 4.3.1.5.5 and noted values for the average F_{en} for charging nozzle and safety injection nozzles are provided in the LRA. During the audit and review, the applicant explained that F_{en} values for normal charging, alternative charging and safety injection nozzles were computed from the actual plant events using an integrated strain rate (ISR) method defined in an EPRI Report (TR-1003083, Guidelines for Assessing Fatigue Environmental Effects in a License Renewal Application.) The staff noted that the ISR method calculates one F_{en} value for one transient pair, though both the charging nozzle and the safety injection nozzle were designed to several thermal transients. In RAI 4.3-3, the staff asked the applicant to justify how one average F_{en} value per nozzle could be used for more than one transient pairs having significant contribution to the CUF.

In the February 21, 2008 letter, the applicant explained how F_{en} value was derived for the charging nozzle and safety injection nozzle. The applicant explained that for the charging nozzle, the available data were used to establish the F_{en} value for the period before monitoring began. Specifically, the applicant indicated that each transient event's ISR F_{en} values were averaged to determine the F_{en} value for the unmonitored period for that specific transient event. Furthermore, the applicant indicated that the effective F_{en} for the charging nozzle was computed for the entire operating period. The applicant explained that for the safety injection nozzle, the F_{en} was calculated using the integrated strain rate method described in response to audit Question 4.3-05. In addition, the applicant stated that only safety injection events were used to calculate F_{en} while additional thermal and pressure transients (such as heatup and cooldown) do not cause measureable additional fatigue usage. The staff reviewed that applicant's response and noted that it explained the term, average F_{en}, provided in the LRA. In addition, the staff noted that average F_{en} is computed from all the transients accrued to date, and that individual transient event F_{en} value is calculated through the ISR method. On the basis that the applicant stated the term average F_{en}, the staff find the applicant's response acceptable.

The staff reviewed LRA Section 4.3.1.5 and noted it states that the environmentally-assisted fatigue on surge line hot leg nozzle, charging nozzle, and safety injection nozzle was evaluated using fatigue monitoring software. The staff concludes that the validity of fatigue monitoring software was not demonstrated by the applicant as no relevant benchmarking data is available to the staff. In RAI 4.3-4, the staff asked the applicant to provide the benchmarking of the software using relevant transient data, proper 3-D model (cylinder to cylinder), and ANSYS, the computer code acceptable for use. The staff also asked the applicant to justify the use of the fatigue monitoring software to update the CUF calculation by using the monitored or projected transient data (cycles) and to discuss the conservatisms in the calculation on a plant specific basis.

In the February 21, 2008 letter, the applicant provided its benchmarking results of the charging nozzle and surge line hot leg nozzle. The applicant provided its response to demonstrate the conservatisms of the software calculated results to a more refined ASME NB-3200 analysis for these two components. The response provides comparison of Max Salt, Max Sn, Max Ke, and CUF values for two transients per nozzle. The response showed that software analysis produce higher than ASME Analysis CUF values for all transients experienced by the nozzles. The staff reviewed this response, and identified several areas where additional information is needed before a determination of the adequacy and acceptability of the response can be made. Therefore, the staff issued RAI 4.3-5 to request this additional information. In a letter dated May 29, 2008, the applicant provided its response to RAI 4.3-5. In this letter, the applicant also committed to implement a fatigue management program that will use six stress components in the stress based fatigue calculation. On the basis that FatiguePro™ which only calculates 1-D virtual stress will not be used, but six stress components will be calculated, the staff finds the applicant's response acceptable.

4.3.1.5.3 UFSAR Supplement

The applicant provided a UFSAR supplement summary description of its TLAA evaluation of the effects of the RCS environment on fatigue life of piping and components in LRA Section A.3.2.1.

On the basis of its review of the UFSAR supplement, the staff concludes that the summary description of the applicant's actions to address the effects of the RCS environment on fatigue life of piping and components is adequate.

4.3.1.5.4 Conclusion

On the basis of its review, as discussed above, the staff concludes that the applicant has demonstrated, pursuant to 10 CFR 54.21(c)(1)(i), that, for effects of the RCS environment on fatigue life of piping and components, the analyses remain valid for the period of extended operation. The applicant also has demonstrated, pursuant to 10 CFR 54.21(c)(1)(iii), that the effects of aging on the intended function(s) will be adequately managed for the period of extended operation.
The staff also determined that the UFSAR supplement contains an appropriate summary description of the TLAA evaluation, as required by 10 CFR 54.21(d).

4.3.1.6 Full Structural Weld Overlays on Pressurizer Spray Nozzles, Safety and Relief Nozzles, and Surge Nozzles

4.3.1.6.1 Summary of Technical Information in the Application

LRA Section 4.3.1.6 summarizes the evaluation of full structural weld overlays on pressurizer spray nozzles, safety and relief nozzles, and surge nozzles for the period of extended operation.

The applicant has installed full structural weld overlays on the Unit 2 pressurizer spray nozzle, pressurizer safety and relief nozzles, and the pressurizer surge nozzle and plans to install them on the corresponding Unit 1 components at the next refueling outage.

The applicant states that fatigue crack growth analyses using ASME Code Section XI methodology demonstrated the fatigue qualification at the structural weld overlay regions. The impact of the addition of structural weld overlay material on the existing primary stress qualification, considering both deadweight and dynamic loadings, was insignificant. The applicant also states that reconciliation of the existing fatigue evaluation for the limiting locations outside the full structural weld overlays demonstrated that the pressurizer nozzles still would meet ASME Code Section III requirements. The transient assumptions for this analysis are consistent with the existing stress analyses and the Fatigue Monitoring Program.

This change does not affect the cycle counting or cycle-based fatigue modules of the Fatigue Monitoring Program; however, evaluation of the effects of the weld overlays on the program's stress-based module for monitoring the CUF of the pressurizer nozzles continues. As an enhancement to the program, prior to the period of extended operation the applicant states it will evaluate the impact of the full structural weld overlays on the stress-based fatigue modules for the pressurizer nozzles and, if the existing module is not conservative, revise it to continue to provide valid results. The 60-year projection of the CUF for the limiting surge nozzle location, using the existing stress-based fatigue module, is 0.00004.

4.3.1.6.2 Staff Evaluation

The staff reviewed LRA Section 4.3.1.6, pursuant to 10 CFR 54.21(c)(1).

The staff noted that the applicant did not specify whether the TLAA full structural weld overlays on pressurizer spray nozzles, safety, and relief nozzles and surge nozzle were demonstrated to be acceptable in accordance to 10 CFR 54.21(c)1(i), 10 CFR 54.21 (c)1(ii), or 10 CFR 54.21 (c)1(iii). Therefore, the staff asked the applicant to provide clarification as well as to provide a summary of this TLAA during the audit.

In its response, the applicant stated that it will amend the VEGP LRA to include the following information at the end of Section 4.3.1.6:

> In summary, the reconciliation of the existing fatigue evaluation that was performed for the limiting locations outside the FSWOL is a TLAA that remains valid for the period of extended operation because the cycles assumed will not be exceeded during 60 years of operation. Therefore, this TLAA has been demonstrated to be acceptable for the period of extended operation in accordance with 10 CFR 54.21(c)(1)(i).

In a letter dated March 20, 2008, the applicant amended the application according to the response above. The staff reviewed the change and noted that the proper code of federal regulation is cited in the LRA. On this basis, the staff finds the applicant's response acceptable.

4.3.1.6.3 UFSAR Supplement

The applicant provided a UFSAR supplement summary description of its TLAA evaluation of full structural weld overlays on pressurizer spray nozzles, safety and relief nozzles, and surge nozzle in LRA Section A.3.2.1. On the basis of its review of the UFSAR supplement, the staff concludes that the summary description of the applicant's actions to address full structural weld overlays on pressurizer spray nozzles, safety and relief nozzles, and surge nozzle is adequate.

4.3.1.6.4 Conclusion

On the basis of its review as discussed above, the staff concludes that the applicant has demonstrated, pursuant to 10 CFR 54.21(c)(1)(i), that the analyses for full structural weld overlays on pressurizer spray nozzles, safety and relief nozzles, and surge nozzle remain valid for the period of extended operation. The staff also determined that the UFSAR supplement contains an appropriate summary description of the TLAA evaluation, as required by 10 CFR 54.21(d).

4.3.1.7 High-Energy Line-Break Postulated Locations Based on Fatigue Cumulative Usage Factor

4.3.1.7.1 Summary of Technical Information in the Application

LRA Section 4.3.1.7 summarizes the evaluation of high-energy line-break postulated locations based on fatigue CUF for the period of extended operation. Postulation criteria for high-energy line breaks described in UFSAR Section 3.6.2.1.1 are from Branch Technical Position MEB 3-1 except for lines with postulated breaks eliminated by leak-before-break analysis. One of the MEB 3-1 criteria for Class 1 piping postulates pipe breaks at any intermediate locations where the CUF exceeds 0.1. The applicant states that the staff has determined that this postulation qualifies as a TLAA.

The applicant also states that the original 40-year design cycles are conservative for 60 years of operation except for the feedwater cycling transient and the loss of charging and loss of letdown transients; therefore, the CLB postulated intermediate break locations are unaffected except possibly the Class 1 piping by these transients.

For the feedwater cycling transient, the only Class 1 components analyzed are the SGs. The transient impacted SG main and auxiliary feedwater nozzles but the attached piping is not Class 1; therefore, the postulation of intermediate break points in the feedwater and auxiliary feedwater systems is unaffected.

The applicant further states that for the charging line and letdown line transients, CUF monitoring of the charging nozzles assures that all Class 1 chemical volume and control system components continue to have a CUF less than 1.0; however, without more this assurance does not necessarily lead to the conclusion that the CUF of Class 1 chemical volume and control system components with a design CUF less than 0.1 will remain below 0.1. For the period of extended operation, the 0.01316 reduced allowable CUF for the charging nozzle accommodates environmental fatigue effects (LRA Section 4.3.1.5.4) for assurance that the CUF for Class 1 charging and letdown line components with a design CUF less than 0.1 will remain below 0.1.

Prior to the period of extended operation, the applicant states it will ensure that the Fatigue Monitoring Program limits adequately maintain below 0.1 the CUF for Class 1 portions of the charging and letdown lines with a design analysis CUF of less than 0.1 consistent with the CLB for postulated intermediate break locations.

4.3.1.7.2 Staff Evaluation

The staff reviewed LRA Section 4.3.1.7, pursuant to 10 CFR 54.21(c)(1).

The staff reviewed the application, and noted that it is not clear whether the TLAA "High Energy Line-Break Postulated Locations Based on fatigue Cumulative Usage Factor" is demonstrated to be acceptable in accordance to 10 CFR 54.21(c)(i), 10 CFR 54.21 (c)(ii), or 10 CFR 54.21 (c)(iii). Therefore, the staff asked the applicant to provide clarification as well as to provide a summary of this TLAA during the audit.

In its response, the applicant stated that after the VEGP LRA was submitted, Westinghouse performed the additional review mentioned in the LRA. The applicant stated that, in accordance with Westinghouse letter GP-18223, the applicant concluded that CUFs at nozzles are higher than those at the adjacent piping components and that if the usage factor calculated for the nozzle subjected to operating transients is less than the design usage factor, it may be concluded that the adjacent class 1 auxiliary piping components' usage factors would be less than their design usage factors if evaluated for the same operating transients. Furthermore, the applicant stated it will amend the VEGP LRA to replace the last two paragraphs of Section 4.3.1.7 with the following information:

> The normal and alternate charging nozzle design usage factors are 0.995. The maximum usage factors in the piping are 0.90 in Section 1, and 0.40 in Section 2. Based on the system design and operation, the actual operating transients in the piping and nozzles will be similar, consistent with the design transients. Therefore, it is reasonable to conclude that fatigue usage factors calculated for the RCL charging nozzles based on operating transients are bounding for the locations in the adjacent class 1 auxiliary piping. It is also reasonable to conclude that the magnitudes of fatigue usage of the various components will be related in a manner similar to those reported in the design reports.
>
> Therefore, if the usage factor calculated for the nozzle subjected to operating transients is less than the design usage factor, it may be concluded that the adjacent class 1

auxiliary piping components' usage factors would be less than their design usage factors if evaluated for the same operating transients. Therefore, the existing HELB analyses for CVCS piping remain valid as long as the Fatigue Monitoring Program maintains the CUF of the charging nozzles less than or equal to 1.0 (see Section 4.3.1.5.4).

In summary, the existing VEGP HELB analyses have been shown to remain valid for the period of extended operation, except for the VEGP HELB analysis for CVCS piping which is maintained valid by the Fatigue Monitoring Program.

Therefore, this TLAA has been demonstrated to be acceptable for the period of extended operation in accordance with 10 CFR 54.21(c)(1)(i) and 10 CFR 54.21(c)(1)(iii).

In a letter dated March 20, 2008, the applicant amended its application according to the response above. The staff reviewed this amendment as well as the response above and noted that 10 CFR 54.21(c)(1)(i) and 10 CFR 54.21(c)(1)(iii) are the applicable regulations cited in the LRA. On this basis, the staff finds the applicant response acceptable.

4.3.1.7.3 UFSAR Supplement

The applicant provided a UFSAR supplement summary description of its TLAA evaluation of high-energy line-break postulated locations based on fatigue cumulative usage factor in LRA Section A.3.2.1. On the basis of its review of the UFSAR supplement, the staff concludes that the summary description of the applicant's actions to address high-energy line-break postulated locations based on fatigue cumulative usage factor is adequate.

4.3.1.7.4 Conclusion

On the basis of its review, as discussed above, the staff concludes that the applicant has demonstrated, pursuant to 10 CFR 54.21(c)(1)(i), that the analyses for high-energy line-break postulated locations based on fatigue CUF remain valid for the period of extended operation. The applicant has also demonstrated pursuant to 10 CFR 54.21(c)(1)(iii), that the effects of aging on the intended function(s) will be adequately managed for the period of extended operation. The staff also determined that the UFSAR supplement contains an appropriate summary description of the TLAA evaluation, as required by 10 CFR 54.21(d).

4.3.2 Fatigue of ASME Non-Class 1 Components

4.3.2.1 Summary of Technical Information in the Application

LRA Section 4.3.2 summarizes the evaluation of fatigue of ASME Code non-Class 1 components for the period of extended operation. The applicant states that in the license renewal evaluation of cracking due to thermal fatigue for ASME Code components outside the reactor coolant pressure boundary (non-Class 1), thermal stresses on piping bound thermal stresses on other system components. The design of ASME Code Section III Classes 2 and 3 piping systems incorporates stress-reduction factors for determining the acceptability of the piping design for thermal stresses. Components within the scope of license renewal designed to ASME B31.1 requirements also incorporate stress-reduction factors based upon an assumed number of thermal expansion cycles. In general, the calculation of the thermal expansion stress (S_A) assumes 7000 full-temperature thermal cycles leading to a stress reduction factor of 1.0 in the stress analyses.

The applicant states it also evaluated the validity of the assumption of 7000 full-temperature thermal cycles for 60 years of plant operation. The assumption was generally conservative and the actual temperature changes experienced by most systems were much less severe than the design full-temperature cycles.

In some cases, the evaluation converted "partial-cycle" transients with an actual temperature change much less severe than that of the design full-temperature cycles to equivalent full-temperature thermal cycles (or, conversely, converted full-temperature cycles to an allowable number of partial-temperature cycles).

The applicant notes that ANSI B31.1 Power Piping Code, 1967 Edition, Section 102.3.2, provides the following equation and methodology for mathematical determination of the number of equivalent full-temperature range changes that result from the number of lesser temperature range changes:

$$N = N_E + r_1^5 N_1 + r_2^5 N_2 + \ldots r_n^5 N_n$$

Where: N = the number of equivalent full-temperature cycles,

N_E = number of cycles at full temperature change for which expansion stress has been calculated,

$N_1, N_2 \ldots N_n$ = number of cycles at lesser temperature changes,

$r_1, r_2 \ldots r_n$ = ratio of lesser temperature cycles to the cycle for which the expansion stress has been calculated.

The results of this evaluation indicate that the 7000 full-temperature thermal cycle assumptions is valid and bounding for 60 years of operation. Therefore, the applicant further states that existing fatigue analyses for the non-ASME Code Class 1 components are valid for the extended term of operation.

4.3.2.2 Staff Evaluation

The staff reviewed LRA Section 4.3.2 to verify, that pursuant to 10 CFR 54.21(c)(1)(i), the analyses remain valid for the period of extended operation.

The staff reviewed LRA Section 4.3.2 and noted that the LRA states "In general, the assumption was conservative and the actual temperature changes experienced by most systems were less severe than the design full-temperature cycles." The staff noted that the LRA did not provide the plant systems which experienced temperature changes greater than those assumed in the initial design and asked the applicant for the list during the audit. The staff also asked the applicant to explain how these plant systems were evaluated. In its response, the applicant stated that no plant systems were identified that experienced temperature changes greater than those assumed in the initial design. Furthermore, the applicant intends to provide clarification to the LRA by removing the words "in general." In a letter dated March 20, 2008, the applicant amended its LRA with the response above. The staff reviewed the response as well as the amendment and noted that clarification was provided by the applicant. On this basis, the staff finds the applicant's response acceptable.

The staff reviewed LRA Section 4.3.2 and noted that the LRA describes evaluations, which converted "partial-cycle" transients with an actual temperature changes to equivalent full-temperature thermal cycles (or conversely, converted full-temperature cycles to an allowable number of partial-temperature cycles). During the audit and review, the staff asked the applicant to identify which systems were evaluated using "partial cycles" and to discuss how the expansion stress range for the partial cycles is evaluated. In its response, the applicant provided a list of systems evaluated by assuming an actual temperature change and determining how many of those "partial cycles" would result in the same thermal expansion stress as 7000 full-temperature cycles using design temperature changes. Furthermore, the applicant will clarify the application by removing the following text "or, conversely, converted full-temperature cycles to an allowable number of partial-temperature cycles." In the March 20, 2008 letter, the applicant submitted an amendment to the LRA according to its response above.

The staff reviewed the applicant's response as well as the calculation that SNC performed for evaluations of partial cycles and noted that clarification was provided by the applicant. On this basis, the staff finds the applicant's response acceptable.

During the audit and review, the staff reviewed the basis document, VEGP-LR-TLAA-307, and noted that it states that the fatigue analysis related to letdown heat exchanger, containment coil and MSIV are determined as TLAA. The staff noted these TLAAs are not discussed in the LRA. Therefore, the staff asked the applicant to provide the details of these TLAA.

In its response, the applicant stated it will add the following paragraphs to the end of LRA Section 4.3.2:

> There are non-Class 1 fatigue evaluations that use a different method of analysis than the 7000 cycles described above. In general, those evaluations use the same cycles, or a subset of the cycles, used for the Class 1 piping and therefore the existing analysis remains valid for 60 years because the cycles assumed will not be exceeded in 60 years.
>
> One case is the analysis that addresses fatigue of the letdown heat exchangers. That analysis utilizes some of the primary piping transient events. The calculation demonstrates that a fatigue exemption applies to the heat exchanger and shows the damage factor for the heat exchanger bolting to be satisfactory with the ring spacer. The cycles assumed for both the heat exchanger and the bolting are bounded by the Class 1 piping cycles. Therefore, this analysis is determined to be a TLAA, but the analysis is already valid for the period of extended operation in accordance with 10 CFR 54.21(c)(1)(i).
>
> Another case is the fatigue test report for containment cooler Copper-Nickel Alloy Cooling Coils. This test report evaluates, by experiment, the stress placed upon Cu-Ni coils due to 1500 thermal cycles over a 40-year design life. The transient cycles that most apply to the cooling coils are those of plant start-up and shutdown (when the containment experiences the greatest temperature change). The limits for RCS start-up and Shutdown (200 of both) will limit the cycles that the coolers see to much less than 1500 for 60 years. Therefore, this analysis is determined to be a TLAA, but the analysis is already valid for the period of extended operation in accordance with 10 CFR 54.21(c)(1)(i).

Another case is the fatigue analysis of the main steam isolation valves that uses the maximum number of cycles in specification AX4AR17 (2000 for 40 years). The calculation shows that for the maximum yoke stress as calculated, 10000 cycles are allowed. This is 5 times the minimum acceptable per the spec. The component fatigue is bounded by the piping fatigue, which is assured through limits on the number of piping cycles in the Fatigue Monitoring Program. Therefore, this analysis is determined to be a TLAA, but the analysis is already valid for the period of extended operation in accordance with 10 CFR 54.21(c)(1)(i).

In the March 20, 2008 letter, the applicant amended the LRA as stated above. The staff found that all three TLAAs have been described in sufficient technical detail. On this basis, the staff finds the applicant's response acceptable.

4.3.2.3 UFSAR Supplement

The applicant provided a UFSAR supplement summary description of its TLAA evaluation of fatigue of ASME non-Class 1 components in LRA Section A.3.2.2. On the basis of its review of the UFSAR supplement, the staff concludes that the summary description of the applicant's actions to address fatigue of ASME non-Class 1 components is adequate.

4.3.2.4 Conclusion

On the basis of its review, as discussed above, the staff concludes that the applicant has demonstrated, pursuant to 10 CFR 54.21(c)(1)(i), that, for fatigue of ASME non-Class 1 components, the analyses remain valid for the period of extended operation. The staff also determined that the UFSAR supplement contains an appropriate summary description of the TLAA evaluation, as required by 10 CFR 54.21(d).

4.3.3 Fatigue of the Reactor Coolant Pump Flywheel

4.3.3.1 Summary of Technical Information in the Application

LRA Section 4.3.3 summarizes the evaluation of fatigue of the reactor coolant pump flywheel for the period of extended operation. A calculation for the reactor coolant pump flywheels assumes that each pump will be subjected to 6000 start-stop cycles over a 60-year life. The Fatigue Monitoring Program limits each reactor coolant pump to 1000 start-stop cycles, and 60-year projections indicate 632 pump starts at the end of the period of extended operation. Therefore, the applicant determined that the 6000 start-stop cycles remain bounding for 60 years of operation and the fatigue evaluation of the reactor coolant pump flywheels is demonstrably valid for the period of extended operation.

4.3.3.2 Staff Evaluation

The staff reviewed LRA Section 4.3.3 to verify pursuant to 10 CFR 54.21(c)(1)(i), that the analyses remain valid for the period of extended operation.

The staff reviewed LRA Section 4.3.3 and noted the 60-year projections for each reactor coolant pump is within the limit set by the Fatigue Monitoring Program.

The staff reviewed the calculations on the reactor coolant pump and confirmed the applicant's statement. On this basis, the staff finds the applicant's cycle projection acceptable.

4.3.3.3 UFSAR Supplement

The applicant provided a UFSAR supplement summary description of its TLAA evaluation of fatigue of the reactor coolant pump flywheel in LRA Section A.3.2.3. On the basis of its review of the UFSAR supplement, the staff concludes that the summary description of the applicant's actions to address fatigue of the reactor coolant pump flywheel is adequate.

4.3.3.4 Conclusion

On the basis of its review, as discussed above, the staff concludes that the applicant has demonstrated, pursuant to 10 CFR 54.21(c)(1)(i), that, for fatigue of the reactor coolant pump flywheel, the analyses remain valid for the period of extended operation. The staff also determined that the UFSAR supplement contains an appropriate summary description of the TLAA evaluation, as required by 10 CFR 54.21(d).

4.3.4 Fatigue of Reactor Vessel Supports

4.3.4.1 Summary of Technical Information in the Application

LRA Section 4.3.4 summarizes the evaluation of fatigue of reactor vessel supports for the period of extended operation. The Westinghouse Generic Technical Report WCAP 14422, Revision 2a, considers fatigue of reactor vessel supports a potential TLAA for supports constructed in accordance with the 1963 version of the American Institute of Steel Construction (AISC) Code. The applicant states that In the SER for this WCAP report, staff indicates that license renewal applicant must ensure a later AISC Code version was used. If not, the applicant should address the issues related to the aging effect from fatigue. The applicant further states that the VEGP design used the AISC Code 1969 version and, therefore, the existing analysis is demonstrably valid for the period of extended operation.

4.3.4.2 Staff Evaluation

The staff reviewed LRA Section 4.3.4 to verify, pursuant to 10 CFR 54.21(c)(1)(i), that the analyses remain valid for the period of extended operation.

During the audit and review, the staff reviewed the basis documents for metal fatigue time-limited aging analyses, VEGP-LR-TLAA-307 and VEGP-LR-TE-018. The staff noted that the applicant's basis documents state that an LRA amendment is required to correct the LRA statement on the design basis of reactor vessel support. The staff asked the applicant to confirm the amendment will revise LRA section 4.3.4 from "Code of Record for VEGP as AISC 1969 version" to "ASME Code Section III. NF in accordance with UFSAR Table 3.2.2-1."

In its response, the applicant stated that the reactor pressure vessel supports embedded within the primary shield are procured in accordance with ASME Code, Section III, Division 1, Subsection NF; however, since they are outside the ASME Jurisdictional boundary, their design follows AISC specifications. Furthermore, the applicant stated both the 1969 version of the AISC Code and ASME Code, Section III, Division 1, Subsection NF apply to the supports.

In a letter dated March 20, 2008, the applicant revised the LRA to reflect that both codes apply to reactor pressure vessel supports. The staff reviewed the applicant's response and its amendment change, and noted that the reactor pressure vessel supports were designed to an

AISC code version later than the AISC 1963 code. On this basis, the staff finds the applicant's response applicable because it is consistent with the SER for WCAP 14422, Revision 2a.

4.3.4.3 UFSAR Supplement

The applicant provided a UFSAR supplement summary description of its TLAA evaluation of fatigue of reactor vessel supports in LRA Section A.3.2.4. On the basis of its review of the UFSAR supplement, the staff concludes that the summary description of the applicant's actions to address fatigue of reactor vessel supports is adequate.

4.3.4.4 Conclusion

On the basis of its review, as discussed above, the staff concludes that the applicant has demonstrated, pursuant to 10 CFR 54.21(c)(1)(i), that, for fatigue of reactor vessel supports, the analyses remain valid for the period of extended operation. The staff also determined that the UFSAR supplement contains an appropriate summary description of the TLAA evaluation, as required by 10 CFR 54.21(d).

4.3.5 Fatigue of Steam Generator Secondary Manway and Handhole Bolts

4.3.5.1 Summary of Technical Information in the Application

LRA Section 4.3.5 summarizes the evaluation of fatigue of SG secondary manway and handhole bolts for the period of extended operation. The applicant states that a Westinghouse calculation for SG secondary manway and handhole bolts assumed the same transients as those for Class 1 component fatigue evaluations and found a qualified life for the manway bolts of only 20 years. In 1993, there was a determination that a low-temperature rerate would reduce the qualified life of the manway bolts to 14.5 years. The applicant also states that a new secondary side manway and handhole bolts fatigue evaluation based on revised cycles applicable to 40 years of operation qualified the bolts for the current operating period with rerating. The 40-year CUF based on revised cycles for 40 years with rerate is 0.997 for secondary side manway bolts and 0.724 for the handhole bolts. The applicant elected to replace secondary manway and handhole bolts at 30 years of service life in lieu of fatigue monitoring.

The applicant further states that it will continue to ensure that the bolt design limit is not exceeded during the period of extended operation by periodically replacing both secondary side manway and handhole bolts under the Bolting Integrity Program, to manage fatigue. The manway bolts have not been replaced previously and are scheduled for replacement at 30 years of service life during the 2017 and 2019 spring outages for Units 1 and 2, respectively. The handhole bolts were replaced in 1996 and 1998 for Units 1 and 2, respectively, and are scheduled to be replaced again during the 2026 and 2028 spring outages for Units 1 and 2, respectively; therefore, the Bolting Integrity Program adequately manages SG secondary manway and handhole bolt fatigue for the period of extended operation.

4.3.5.2 Staff Evaluation

The staff reviewed LRA Section 4.3.5, to verify pursuant to 10 CFR 54.21(c)(1)(iii), that the effects of aging on the intended function(s) will be adequately managed for the period of extended operation.

The staff noted the applicant proposed to periodically replace secondary manway and handhole bolts at 30 years of service life in its application. This information contradicts the Vogtle License Renewal Commitment List, which states that this replacement schedule may be changed due to an updated analyses initiated by the Bolting Integrity Program. During the audit, the staff asked the applicant to explain why 10 CFR 54.21(c)(1)(iii) is used if these bolts' replacement schedule is dependent on future analyses.

In its response, the applicant indicated there is currently no plan to perform an analysis under the Bolting Integrity Program that could be used to change the scheduled replacement of steam generator manway and handhole bolts. Furthermore, the applicant stated that, should SNC decide at a later date to revise the replacement schedule of these bolts, an analysis would be performed to justify the revision.

The staff reviewed the applicant's response, and noted the replacement schedule for steam generator manway and handhole bolts is based on a potential replacement schedule. The staff also noted that this potential is still part of the Bolting Integrity Program. Therefore, these bolts are still managed under the Bolting Integrity Program and 10 CFR 54.21(c)(1)(iii) still applies. On this basis, the staff finds the applicant's response acceptable. In addition, the staff noted that the applicant amended Commitment # 30 (on replacement schedule for steam generator manway and handhold bots) to reflect the above change. The staff reviewed the change and noted that Commitment # 30 is now consistent with the applicant response and therefore acceptable.

4.3.5.3 UFSAR Supplement

The applicant provided a UFSAR supplement summary description of its TLAA evaluation of fatigue of SG secondary manway and handhole bolts in LRA Section A.3.2.5. On the basis of its review of the UFSAR supplement, the staff concludes that the summary description of the applicant's actions to address fatigue of SG secondary manway and handhole bolts is adequate.

4.3.5.4 Conclusion

On the basis of its review, as discussed above, the staff concludes that the applicant has demonstrated, pursuant to 10 CFR 54.21(c)(1)(iii), that, for fatigue of SG secondary manway and handhole bolts, the effects of aging on the intended function(s) will be adequately managed for the period of extended operation.

The staff also determined that the UFSAR supplement contains an appropriate summary description of the TLAA evaluation, as required by 10 CFR 54.21(d).

4.3.6 Fatigue of Reactor Vessel Internals

4.3.6.1 Summary of Technical Information in the Application

LRA Section 4.3.6 summarizes the evaluation of fatigue of reactor vessel internals for the period of extended operation. The applicant states that the original design did not require a fatigue analysis of the reactor vessel internals; however, for rerating, a Westinghouse calculation for reactor vessel internals that assumed the same cycles as those for Class 1 component fatigue evaluations found CUFs less than 1.0 for all subcomponents evaluated. As the analysis utilized the same design transients as those for the Class 1 component evaluations, the evaluation of the ASME Class 1 piping and component design transient cycles applies also to the reactor vessel internals.

The applicant further states that the design cycles for the transients applicable to the reactor vessel internals were conservative, therefore; the reactor vessel fatigue evaluation remains valid for the period of extended operation.

4.3.6.2 Staff Evaluation

The staff reviewed LRA Section 4.3.6 to verify pursuant to 10 CFR 54.21(c)(1)(i), that the analyses remain valid for the period of extended operation.

The staff reviewed the LRA Section 4.3.6 and noted that the LRA states the design cycles for the transient applicable to the reactor vessel internals were shown to be conservative. During the audit and review, the staff asked the applicant to justify why the current design cycles for reactor vessel internals is conservative considering some of the transients are projected to exceed design cycles.

In its response, the applicant stated that:

> The only events that are projected to exceed design cycles at VEGP are feedwater cycling, letdown trip delayed and charging trip prompt. Feedwater cycling event provides a negligible contribution to the fatigue of the reactor vessel internals. The other two events are auxiliary transients and they are not considered in the fatigue evaluation of the reactor vessel.

The staff reviewed the applicant's response and noted that the only applicable transient that is projected to exceed the design cycle is the feedwater cycling transient that provides only a negligible contribution to the fatigue of the reactor vessel internals. The staff reviewed the LRA and noted all other transient cycles are within their design limits, and therefore the design cycles for the transients applicable to the reactor vessel internals can be considered conservative. On this basis, the staff finds the applicant's response acceptable.

4.3.6.3 UFSAR Supplement

The applicant provided a UFSAR supplement summary description of its TLAA evaluation of fatigue of reactor vessel internals in LRA Section A.3.2.6. On the basis of its review of the UFSAR supplement, the staff concludes that the summary description of the applicant's actions to address fatigue of reactor vessel internals is adequate.

4.3.6.4 Conclusion

On the basis of its review, as discussed above, the staff concludes that the applicant has demonstrated, pursuant to 10 CFR 54.21(c)(1)(i), that, for fatigue of reactor vessel internals, the analyses remain valid for the period of extended operation. The staff also determined that the UFSAR supplement contains an appropriate summary description of the TLAA evaluation, as required by 10 CFR 54.21(d).

4.4 Environmental Qualification of Equipment

The 10 CFR 50.49 EQ program is a TLAA for purposes of license renewal. The TLAA of the EQ electrical components includes all long-lived, passive, and active electrical and I&C components that are important to safety and located in a harsh environment. The harsh environments of the plant are those areas subject to environmental effects by loss of coolant accidents or high-energy line breaks. EQ equipment comprises safety-related and Q-list equipment, nonsafety-related equipment the failure of which could prevent satisfactory accomplishment of any safety-related function, and necessary post-accident monitoring equipment.

As required by 10 CFR 54.21(c)(1), the applicant must provide a list of EQ TLAAs in the LRA. The applicant shall demonstrate that for each type of EQ equipment, one of the following is true: (1) the analyses remain valid for the period of extended operation, (2) the analyses have been projected to the end of the period of extended operation, or (3) the effects of aging on the intended function(s) will be adequately managed for the period of extended operation.

4.4.1 Summary of Technical Information in the Application

LRA Section 4.4 summarizes the evaluation of environmental qualification (EQ) of equipment for the period of extended operation. Section 50.49 and Appendix A of 10 CFR Part 50 establish EQ requirements. The applicant established its EQ program to demonstrate that certain electrical components are qualified to perform safety functions in the harsh environment following a design-basis accident. Elements of the proof of qualification involve the original 40-year license period; hence, EQ Program qualification reports and calculations meet the TLAA definition. In general, the applicant did not establish for EQ Program components qualified lives longer than the original 40-year license period.

EQ of mechanical equipment is through stringent selection of materials for use under adverse environmental conditions supported by partial type testing and material analysis and evaluation to confirm the adequacy of the materials.

As a result of this application, there will be no components added to the EQ Program. Qualified EQ component service lives already determined are tracked to determine when a component nears the end of its service life. For components nearing the end the EQ Program re-evaluates them for longer service refurbished or requalified or for replacement. The EQ Program is procedurally in compliance with 10 CFR Part 50, Appendix B, and routinely audited for quality assurance. This program will be continued through the period of extended operation; hence, the TLAAs will be managed by an AMP in accordance with 10 CFR 54.21(c)(1)(iii).

The program discussion addressees the EQ component reanalysis attributes (analytical methods, data collection and reduction methods, underlying assumptions, acceptance criteria, and corrective action).

4.4.2 Staff Evaluation

The staff reviewed LRA Section 4.4, to verify pursuant to 10 CFR 54.21(c)(1)(iii), that the effects of aging on the intended function(s) will be adequately managed for the period of extended operation.

The staff reviewed LRA Section 4.4 and the program basis document to determine whether the applicant had submitted adequate information to meet the requirement of 10 CFR 54.21(c)(1). For the electrical equipment identified in LRA Table 4.1-1, the applicant used 10 CFR 54.21(c)(1)(iii) to demonstrate that the aging effects of EQ equipment will be adequately managed during the period of extended operation. The staff reviewed the Environmental

Qualification Program to determine whether it will ensure that the electrical and I&C components covered under this program will continue to perform their intended functions consistent with the CLB for the period of extended operation. The staff's evaluation of the components' qualification focused on how the Environmental Qualification Program manages the aging effects to meet the requirements delineated in 10 CFR 50.49.

The staff conducted an audit of the information provided in LRA Section B.3.37 and program bases documents. The staff's evaluation is documented in SER Section 3.0.3.1.8. On the basis of its audit, the staff finds that the Environmental Qualification Program, which the applicant claimed to be consistent with the GALL AMP X.E1, "Environment Qualification of Electrical Components," is consistent with the GALL Report.

Therefore, the staff finds that the program is capable of programmatically managing the qualified life of components within the scope of the program for license renewal.

4.4.3 UFSAR Supplement

The applicant provided a UFSAR supplement summary description of its TLAA evaluation of environmental qualification of equipment in LRA Section A.3.3. On the basis of its review of the UFSAR supplement, the staff concludes that the summary description of the applicant's actions to address environmental qualification of equipment is adequate.

4.4.4 Conclusion

On the basis of its review, as discussed above, the staff concludes that the applicant has demonstrated, pursuant to 10 CFR 54.21(c)(1)(iii), that, for environmental qualification of equipment, the effects of aging on the intended function(s) will be adequately managed for the period of extended operation. The staff also determined that the UFSAR supplement contains an appropriate summary description of the TLAA evaluation, as required by 10 CFR 54.21(d).

4.5 Concrete Containment Tendon Prestress

4.5.1 Summary of Technical Information in the Application

In VEGP LRA Section 4.5, the applicant summarizes the evaluation of concrete containment tendon prestress for the period of extended operation. The containment structure consists of a prestressed reinforced concrete cylinder and hemispherical dome. Containment prestressing is by a two-way post-tensioning system consisting of circumferential hoop tendons and two groups of inverted U-shaped tendons. The prestressing tendons in prestressed concrete containments lose their prestressing forces with time due to creep and shrinkage of the concrete and relaxation of the prestressing steel. Loss of tendon prestressing force is a TLAA; therefore, the adequacy of the prestressing forces is evaluated for the period of extended operation.

The applicant states that as part of tendon surveillance, under the IWL Inservice Inspection Program, a sample of tendons from each group (inverted U-shaped and hoop) in each inspection interval is inspected to ensure that the acceptance criteria are met and tendon prestressing forces will remain above the minimum required values for the succeeding inspection intervals. The program also recalculates the regression analysis trend lines of these two tendon groups from individual tendon forces consistent with Information Notice 99-10 (i.e., using individual tendon data rather than averages) to verify if the average prestressing force in each tendon group will remain above their minimum required values for the remainder of the licensed operating period.

The applicant also states that when extended to 60 years, the regression analysis demonstrated that prestress in both the inverted U-shaped and horizontal ("hoop") tendon groups should remain above the applicable minimum required values for at least 60 years of operation and that all tendons, therefore, should maintain their design-basis function for the period of extended operation without retensioning. Similarly, no individual tendon data from the "common tendons" (one inverted U-shaped and one horizontal with prestress measured at each surveillance) or from the other sample tendons tested, to date, show a loss of prestress, sufficient to indicate a possible need to retension during the period of extended operation.

The applicant further states that consistent with 10 CFR 54.21(c) (1) Option (ii), acceptable prestress forces in containment tendons have been projected to the end of the period of extended operation.

4.5.2 Staff Evaluation

The staff reviewed LRA Section 4.5 according to the requirements of 10 CFR 54.21(c)(1)(ii) to verify that (1) the trend of prestressing forces in each tendon group has been projected to the end of the period of extended operation and (2) the projected prestressing forces are above their respective minimum required values. The following discussion describes the staff's RAI related to the LRA Section 4.5, the corresponding applicant responses, and the staff evaluation.

In Figure 4.5-4a, "Unit 2 Vertical Baseline Tendon V20-92," and Figure 4.5-5a, "Unit 2 Shell Hoop Baseline Tendon H-99," of the LRA, the tendon force in the seventh year is larger than that in the third year. By letter dated January 28, 2008, the staff issued RAI 4.5-1 to request that the applicant discuss this anomaly and determine its effect on the regression analysis and the corresponding trend lines.

By letter dated February 27, 2008, the applicant responded to RAI 4.5-1 and stated that the reason that the tendon force in the seventh year is larger than that in the third year can be attributed to the method of lift-off measurement performed by two different vendors for the third and seventh years. The applicant further investigated the trend lines ignoring 3rd year data and then ignoring 7th year data only and, in both cases, 60 year trend line values were greater than the minimum required prestressing force values.

The staff finds the applicant's response to RAI 4.5-1 acceptable because further investigation of the trend lines showed that the 60 year trend line prestressing forces will still be greater than the minimum required values.

In a subsequent telephone conference as summarized in a letter from D. J. Ashley (NRC) to Southern Nuclear Operating Company dated March 26, 2008, the staff noted that all figures depicting the tendon lift-off force trend lines in Section 4.5 are based on years after the initial Structural Integrity Test (SIT) which may not coincide with the years after the initial date of operation. As such, the staff requested the applicant to reconcile the difference between the initial SIT and the initial date of operation.

The applicant stated that the difference between the initial SIT and the initial date of operation is approximately 4 months for each unit at VEGP and the effects of this 4 month difference on the results of the regression analysis, as shown in Section 4.5 of the LRA will be insignificant.

Because of the logarithmic scale of time in Figures 4.5-1 through 4.5-6, included in Section 4.5 and margins available between the 60 year trend values and the minimum required values of prestressing shown in Table 4.5-7 of the LRA, the staff concurs that a 4 month difference will not significantly affect the trend line and the prestress in the inverted U-shaped and hoop tendon groups will remain above the applicable minimum required values for 60 years of operation. The staff recognizes that because of subsequent containment tendon surveillance lift-off results, the trend line presently shown in Section 4.5 of the LRA could change. The staff accepts the applicant's assertion that this attribute will be managed by the containment in-service inspection program which entails taking appropriate corrective actions, should the projected tendon force for a tendon group fall below the minimum required value.

4.5.3 UFSAR Supplement

The applicant provided a UFSAR supplement summary description of its TLAA evaluation of concrete containment tendon prestress in LRA Section A.3.4. On the basis of its review of the UFSAR supplement, the staff concludes that the summary description of the applicant's actions to address concrete containment tendon prestress is adequate.

4.5.4 Conclusion

On the basis of its review, as discussed above, the staff concludes that the applicant has demonstrated pursuant to 10 CFR 54.21(c)(1)(ii), that (1) for the TLAA for concrete containment prestressing tendons, the trend of prestressing force in each tendon group has been projected to the end of the period of extended operation; and (2) the projected prestressing force for each tendon group remains above its respective minimum required value. The staff also determined that the UFSAR supplement contains an appropriate summary description of the TLAA on containment tendon loss of prestress analysis for the period of extended operation, as required by 10 CFR 54.21(d).

4.6 Penetration Load Cycles

4.6.1 Summary of Technical Information in the Application

LRA Section 4.6 summarizes the evaluation of penetration load cycles for the period of extended operation. Fatigue evaluations were required for some of the containment penetrations and for analyses qualifying as TLAAs. The applicant compared the transient assumptions for those evaluations against the transient assumptions for Class 1 component fatigue and determined that the penetration fatigue evaluations are adequate and require no updating for license renewal.

The number of transients assumed for the penetrations is bounded by the number of transients assumed for Class 1 piping. Per LRA Section 4.3.1, the 40-year design cycles are maintained as the 60-year design cycles except that the applicant credits CUF monitoring of the main and auxiliary feedwater nozzles rather than monitoring of feedwater cycling events and credits CUF monitoring of the charging nozzles rather than monitoring of loss of charging and loss of letdown events.

The penetration fatigue calculations do not assume the number of feedwater cycles. Only one penetration fatigue calculation considers the number of loss of charging or loss of letdown events in calculating CUF, and the applicant has determined that none of the cycles assumed in that fatigue evaluation will be exceeded within the period of extended operation.

4.6.2 Staff Evaluation

The staff reviewed LRA Section 4.6 to verify pursuant to 10 CFR 54.21(c)(1)(i), that the analyses remain valid for the period of extended operation.

During the audit and review, the staff reviewed the applicant's specification "Specification for Pipe Penetrations for Georgia Power Company Alvin W. Vogtle Nuclear Plant Burke County, Georgia," November 12, 1987, (X4AQ10, Rev.7) to identify the transient assumptions used for the fatigue evaluation of the penetrations. The staff's review of the transient assumptions showed that all of them were bounded by the VEGP USFAR Table 3.9.N.1-1 transients, except for Penetrations 1, 2, 3, 4, 7, 8, 9, 10, 11A, 11B, 12B, & 12C, the applicant assumed 2000 feedwater cycles and for Penetration 4, the applicant assumed loss of charging and loss of letdown. Since the transient assumptions for all the penetrations, except those listed above, were bounded by the cycle limits assumed in the Fatigue Monitoring Program, the existing analyses remain valid for those penetrations.

The staff noted that the fatigue evaluation AX4AQ10-129-1, dated September 26, 1986, calculated the fatigue at a number of nodes. While the alternating stress intensity and the associated temperature was different for each, the cycles used/allowed and the calculated fatigue usage was the same for all transients. Specifically, for each node, two event pairs evaluated were some combination of steady state at 290, a ramp up to 430, a ramp down to 290, and steady state at 290. In each case, the fatigue calculation for each event pair was:

$$\text{Cycles used/allowed} = 185,680/1,000,000 = 0.18568 \text{ (Partial Usage)}$$

Adding the calculated partial usage for each event pair resulted in the cumulative fatigue usage (CUF) for each node evaluated is 0.37136.

Therefore, it can be seen that the fatigue calculation assumed a total of 185,680 cycles, 880 of which are attributable to the loss of charging/loss of letdown events. The resulting CUF is 0.37136. Doubling the assumed number of loss of charging/loss of letdown events to 400 would increase the number in the calculation from 880 of 185,680 to 1760 of 186,560 and increase the calculated CUF to:

$$186,560/1,000,000 = 0.18656 \text{ (Partial Usage for first event pair)}$$

$$186,560/1,000,000 = 0.18656 \text{ (partial Usage for second event pair)}$$

$$\text{Or} \quad 0.37312 \text{ (Total CUF)}$$

For the penetrations that assumed 2000 feedwater cycles, the transient is considered in determining if a fatigue calculation is required and the resulting stresses are used in the calculation for determining the allowed cycles. However, the resulting temperature cycles are only 31.5 °F, which results in a multiplication factor of 0 and the effective number of ΔT cycles of 0. On the above basis, the staff finds, that penetration fatigue calculation also remains valid through the period of extended operation in accordance with 10 CFR 54.21(c)(1)(i). The staff finds it acceptable because the feedwater containment penetrations are not subject to a temperature range that would be high enough to induce thermal stresses in the components.

4.6.3 UFSAR Supplement

The applicant provided a UFSAR supplement summary description of its TLAA evaluation of penetration load cycles in LRA Section A.3.5. On the basis of its review of the UFSAR supplement, the staff concludes that the summary description of the applicant's actions to address penetration load cycles is adequate.

4.6.4 Conclusion

On the basis of its review, as discussed above, the staff concludes that the applicant has demonstrated, pursuant to 10 CFR 54.21(c)(1)(i), that, for penetration load cycles, the analyses remain valid for the period of extended operation. The staff also determined that the UFSAR supplement contains an appropriate summary description of the TLAA evaluation, as required by 10 CFR 54.21(d).

4.7 Other Plant Specific Analysis

4.7.1 Leak-Before-Break Analysis

4.7.1.1 Summary of Technical Information in the Application

LRA Section 4.7.1 summarizes the evaluation of leak-before-break (LBB) analysis for the period of extended operation. UFSAR Section 3.6.1 summarizes plant-specific LBB analyses for both units. These analyses technically justify the elimination of postulated breaks in the reactor coolant loop piping (except for the Unit 1 accumulator and residual heat removal (RHR) branch connections) and pressurizer surge line from the structural design basis.

The applicant states that the analyses for those LBB applications were performed by Westinghouse and the applicant's submittals were accepted by NRC staff. NUREG-1137, the Safety Evaluation Report for VEGP, evaluated the reactor coolant loop analysis (WCAP-10551) and incorporated it into the original final safety analysis report. Through NUREG-1137 Supplement 7 issued in January 1988, the staff accepted the methodology and the results of WCAP-11531, -11583, and -11589 submitted for review.

The aging effect addressed in these evaluations is cracking, specifically in LBB crack stability evaluations for enveloping critical locations determined from loading, pipe geometry, and fracture toughness considerations. A fatigue crack growth analysis also demonstrated negligible fatigue crack growth. Assumptions in these analyses with potential bases in the original 40-year term of operation are the fracture toughness properties for cast austenitic stainless steel (CASS) materials (due to thermal aging) and the design transients cumulative cycles.

The applicant also states that its TLAA evaluation of the primary loop analyses determined that no updates of the pressurizer surge line and Unit 2 RHR line LBB analyses were required for license renewal as these have no CASS materials and as the transients assumed for 40 years are bounding for 60 years (LRA Section 4.3). The applicant determined that the reactor coolant loop and the Unit 2 accumulator line analyses should be updated to account for the extended term because CASS materials are present.

The applicant further states that at its request Westinghouse created an addendum to WCAP-10551 analyzing the reactor coolant loop to account for the addition of thermal aging of the

CASS materials for the period of extended operation. In December 2005 Westinghouse issued Addendum 1, a Westinghouse proprietary Class 2 document, accounting for the effects of thermal aging degradation of the CASS materials over 60 years of operation. Using faulted loads, the analysis demonstrates a margin of at least 2 between the critical flaw and the flaw with a leak rate of 10 gallons per minute (the detectable leakage flow) and a margin of 10 between the calculated leak rate from the detectable leakage locations and the stable detectable leakage flow. The 1.4 margin on loads required by action Item 10 of the final safety evaluation report for WCAP-14575-A is satisfied. No CASS material for Unit 1 and 2 primary loop piping having been replaced. The second part of action Item 10 does not apply.

At the applicant's request Westinghouse created an addendum to WCAP-11583 analyzing the Unit 2 accumulator line to account for the addition of thermal aging of the CASS materials for the period of extended operation and in July 2006 issued Addendum 1, a Westinghouse Proprietary Class 2 document accounting for the effects of thermal aging degradation of the CASS materials over 60 years of operation. The applicant states that using faulted loads; the analysis demonstrates a margin of at least 2 between the critical flaw and the flaw having a leak rate of 10 gallons per minute (the detectable leakage flow) and a margin of 10 between the calculated leak rate from the detectable leakage locations and the stable detectable leakage flow. The 1.4 margin on loads required by action Item 10 of the final safety evaluation report for WCAP-14575-A is satisfied. No CASS material for the Unit 2 accumulator line having been replaced, the second part of action Item 10 does not apply.

The applicant determined that the evaluations for the Unit 2 RHR line and pressurizer surge line were acceptable for the extended license term. The evaluations for the primary coolant loops and the Unit 2 accumulator line were evaluated and updated to address operation through 60 years.

4.7.1.2 Staff Evaluation

The staff reviewed LRA Section 4.7.1, to verify pursuant to 10 CFR 54.21(c)(1)(i), that the analyses remain valid for the period of extended operation and, pursuant to 10 CFR 54.21(c)(1)(ii), that the analyses have been projected to the end of the period of extended operation.

The staff noted that LBB analyses should address material susceptibility to corrosion, the potential for high residual stresses, and environmental conditions that could lead to degradation by stress corrosion cracking considering 60 years of extended operation. Specifically, Standard Review Plan (SRP) Section 3.6.3 states that primary water stress corrosion (PWSCC) is considered to be an active degradation mechanism in alloy 600/82/182 materials in pressurized water reactor plants.

During the audit, the applicant stated that the alloy 600/82/182 materials within the scope of these analyses are the Alloy 82/182 welds on the primary loop piping and at the pressurizer surge line connection to the pressurizer nozzle. In addition, the applicant stated that as a part of the license renewal program, WCAP-10551-P, Addendum 1, performed a LBB evaluation for the Units 1 and 2 primary loop piping that explicitly addressed the PWSCC concern for the Alloy 82/182 welds in this piping.

For the Alloy 82/182 welds in the pressurizer surge line, the applicant states Full-Structural Weld Overlays (FSWOL) are being implemented under the existing 10 CFR 50 processes using

Alloy 52/152 weld material, eliminating reliance on the Alloy 82/182 welds to mitigate the PWSCC concern.

In a letter dated March 20, 2008, the applicant stated SNC will verify the LBB evaluation in WCAP-10551-p, Addendum 1 meets the conditions of that process or have it re-performed using the acceptable process.

The staff reviewed the applicant's response as well as its Commitment No. 36, and noted that the applicant is committed to address the PWSCC implication on the current LBB analysis in accordance with the future guidance the staff will develop prior to the period of extended operation. On this basis, the staff finds the applicant's response and its LRA amendment acceptable.

The staff reviewed section 4.7.1 of the LRA and noted that it states "For license renewal, SNC performed a TLAA evaluation of the primary loop analyses: SNC determined that no updates of the pressurizer surge line and Unit 2 RHR line LBB analyses were required for license renewal." During the audit and review, the staff reviewed SNC letter NL-06-2768 dated January 3, 2007, that contained a request for NRC approval of a proposed alternative for application of pressurizer nozzle FWSOL on VEGP Unit 2. The staff noted that in this letter the applicant stated that the original leak-before-break (LBB) analyses will be confirmed to be valid after the weld overlays are applied, the amount of shrinkage is determined, and the shrinkage is determined, and the shrinkage stresses are calculated. During the audit and review, the staff noted that it is not clear whether the LBB analyses is to be updated or whether updating analyses will affect the validity of FSOWL during the period of extended operation and asked the applicant to clarify.

In its response, the applicant stated that the existing LBB reports were reviewed and determined to remain valid for the period of extended operation because the surge line contains no CASS material and the transient assumptions used in the fatigue portion of that evaluation remain valid for 60 years operation. In addition, the applicant pointed out that the implementation of the FSWOLs is accomplished under the existing 10 CFR 50 process, which meant that the period of extended operation will not impact the validity of the LBB analyses performed for the FSWOL effort.

The staff reviewed the applicant's response and noted that review of LBB analyses has been completed and determined to remain valid for the period of extended operation by Westinghouse. On this basis, the staff finds the applicant's response acceptable.

4.7.1.3 UFSAR Supplement

The applicant provided a UFSAR supplement summary description of its TLAA evaluation of leak-before-break analysis in LRA Section A.3.6.1. On the basis of its review of the UFSAR supplement, the staff concludes that the summary description of the applicant's actions to address leak-before-break analysis is adequate.

4.7.1.4 Conclusion

On the basis of its review, as discussed above, the staff concludes that the applicant has demonstrated, pursuant to 10 CFR 54.21(c)(1)(i), that, for leak-before-break analysis, the analyses remain valid for the period of extended operation. The applicant also has demonstrated, pursuant to 10 CFR 54.21(c)(1)(ii), that the analyses have been projected to the end of the period of extended operation. The staff also determined that the UFSAR supplement contains an appropriate summary description of the TLAA evaluation, as required by 10 CFR 54.21(d).

4.7.2 Fuel Oil Storage Tank Corrosion Allowance

4.7.2.1 Summary of Technical Information in the Application

LRA Section 4.7.2 summarizes the evaluation of fuel oil storage tank corrosion allowance for the period of extended operation. UFSAR Section 9.5.4.2.1.1 states that the diesel fuel oil storage tanks, with no cathodic protection, have a liberal corrosion allowance of 1/8 inch. The applicant discussed the fuel oil storage tank corrosion allowance with the staff. The response to SER Open Item 10 for the UFSAR proves that there is sufficient corrosion allowance in the buried fuel oil storage tanks and piping for an assumed failure size in the coatings of the equipment. The response uses 40 years as the duration for the corrosion allowance. A substitution of 60 years into the calculations indicates that the corrosion allowance should not be exceeded during the period of extended operation of the plant. In this case, the soils analysis shows that the piping will not be affected adversely by corrosion (from the outside into the tank).

For the period of extended operation, the Buried Piping and Tanks Aging Management Program will manage loss of material due to corrosion on the tank exterior (exposed to soil).

The Buried Piping and Tanks Aging Management Program addresses the aging effect of corrosion for these components demonstrably in accordance with 10 CFR 54.21(c)(1)(iii).

4.7.2.2 Staff Evaluation

The staff reviewed LRA Section 4.7.2, to verify pursuant to 10 CFR 54.21(c)(1)(iii), that the effects of aging on the intended function(s) will be adequately managed for the period of extended operation.

10 CFR 54.21(c)(1) allows for TLAAs to be accepted in accordance with one of the following three options for TLAA acceptance:

  (i)   the analysis remains acceptable for the period of extended operation

  (ii)   the analysis has been projected through the expiration of the period of extended operation

  (iii) the effects of aging on the intended function(s) will be managed for the period of extended operation

The applicant stated that the effects of corrosion on the intended structural integrity function of the fuel oil storage tank will be managed for the period of extended operation in accordance with the provisions of 10 CFR 54.21(c)(1)(iii), and credited AMP B.3.4, Buried Piping and Tanks Inspection Program as the program for managing loss of material or cracking due to corrosion in the fuel oil storage tank pressure retaining walls.

The staff reviewed the applicant's license renewal basis evaluation document for the fuel oil storage tank corrosion allowance analysis to determine which VEGP document or record contained the corrosion allowance analysis for the fuel oil storage tank and to verify that analysis was an analysis that met the six criteria of a TLAA, as defined in 10 CFR 54.3. The staff also reviewed the program attributes for the LRA AMP B.3.4, Buried Piping and Tanks Aging Management Program, as provided in the license renewal basis evaluation document for the AMP, to determine whether this AMP was a valid AMP to credit for aging management under the acceptance criterion of 10 CFR 54.21(c)(1)(iii) and whether the AMP would be capable of managing any loss of material or cracking that may occur in the pressure retaining walls of the tanks prior to a loss of intended function.

The staff concludes that neither the UFSAR for VEGP nor the license renewal basis evaluation document identified the VEGP record that contained the corrosion allowance analysis for the fuel oil storage tank. The staff asked the applicant to: (1) identify which VEGP record specifically contained the corrosion analysis for the fuel oil storage tanks, (2) clarify why the analysis meets the criteria for a TLAA as defined in 10 CFR 54.3, and (3) to clarify what the time parameter and what the acceptance criterion are for this analysis.

The applicant responded to the staff's question in a letter dated February 8, 2008. In its response, the applicant stated that the corrosion allowance evaluation for the fuel oil storage tanks was docketed in a letter to the NRC dated October 2, 1985, and that this evaluation was docketed to resolve Open Item 10 that was issued as part of the NRC's original current licensing basis criteria for the VEGP units, as documented in NUREG 1137, Supplement 1, "Safety Evaluation Report Related to the Operation of Vogtle Electric Generating Plant, Units 1 and 2 [October 1985]." The applicant also stated that this analysis meets all criteria for a TLAA in 10 CFR 54.3 and that the number of remaining plant years, as projected through 60 years of licensed plant operations, was the time parameter for the analysis, and that the analysis allows for 1/8 inch corrosion through the tank wall thickness. This is equivalent to the additional thickness of the corrosion allowance protection that was incorporated into the tanks nominal wall thickness design when the tanks were fabricated. The staff finds the applicant's response to be acceptable because it clarified what the CLB documents and parameters are for the fuel oil storage tank corrosion analysis.

The staff reviewed the fuel oil storage tank corrosion allowance analysis (identified as "SER Open Item 10: Cathodic Protection of Fuel Oil Tanks" in the VEGP letter and report of October 2, 1985) and determined the analysis assessed the remaining amount of corrosion allowance that was incorporated into the design wall thickness of the fuel oil storage tanks and their associated piping. The staff concludes that the analysis used a galvanic voltage and current-based evaluation to assess and quantify the amount of galvanically-induced corrosion that could potentially penetrate into the corrosion allowance for the fuel oil storage tanks and their associated fuel oil delivery piping over a 40-year licensed operational plant life. The staff concludes that the report assumes that galvanically-induced corrosion could penetrate entirely through the entire thickness of the additional corrosion allowance for the fuel oil storage tanks and associated piping, but sets a limit that no galvanically-induced corrosion can penetrate into the nominal design wall thicknesses for these components. The amount of corrosion allowance is 0.00318 m (1/8 inch) for the fuel oil storage tanks and 0.00159 m (1/16 inch) for the associated piping. Table 4.7.2-1 summarizes the results of the analysis:

Table 4.7.2-1: Summary of the Corrosion Allowance Analysis for the Diesel Fuel Oil Storage Tanks and Associated Fuel Oil Delivery Piping

Component	Corrosion Allowance Thickness [1]	Result's of VEGP's 40-Year Licensed Analysis in the CLB (in % Penetration of Allowable Corrosion Allowance Thickness)	Staff's Independent 60-Year Projected Analysis [2] (in % Penetration of Allowable Corrosion Allowance Thickness)	Acceptability Determination [3] (Acceptable: [Yes/No] / Acceptance Criterion
Fuel Oil Storage Tanks	0.00318 meters (1/8 inch)	25%	37.6%	Yes : 10 CFR 54.21(c)(1)(ii)
Associated Fuel Oil Delivery Piping	0.00159 meters (1/16 inch)	50 %	75.5%	Yes : 10 CFR 54.21(c)(1)(ii)

Notes:
1. The report assumes that the entire corrosion allowance can be penetrated (i.e. 100% allowable penetration by galvanic corrosion), but that the penetration cannot extend into the nominal design wall thickness. The report also assumes that the flaw is a 1/32 inch diameter circular pit (flaw) that induces an oxidation cell below the protective coating. The numbers in this column represent the amount of additional corrosion allowance thickness that was incorporated into the component design.

2. The applicant did not include any 60-year projection of the analysis in the LRA but rather opted to manage the aging effect for the period of extended operation in accordance with 10 CFR 54.21(c)(1)(iii). The staff did an independent 60-year calculation of the remaining corrosion allowance based on the applicant's galvanic corrosion methodology to check whether there would be acceptable corrosion allowance even if the applicant did not manage that aging effect for the period of extended operation. The applicant provided its updated 60-year corrosion allowance projections in the applicant's letter of March 20, 2008.

3. The analysis is acceptable in accordance with the provisions of 10 CFR 54.21(c)(1)(ii) if the projected 60-year depth of penetration is less than 100 %, as based on the staff's independent calculation. Alternatively, the applicant may accept this TLAA in accordance with 10 CFR 54.21(c)(1)(iii) if it demonstrates that the effects of galvanic corrosion on the intended structural integrity function of the fuel oil storage tanks and associated delivery piping will be managed for the period of extended operation.

Footnote 3. in Table 4.7.2-1 provides the basis for accepting this TLAA in accordance with the provision in 10 CFR 54.21(c)(1)(ii). The results of the staff's independent 60-year projection of the corrosion allowance demonstrates that the fuel oil storage tanks will have 72.4% remaining margin on the corrosion allowance that was incorporated into the tank design and that the associated fuel oil delivery piping will have a 24.5% remaining margin on the corrosion allowance that was incorporated into the pipe design. This is in addition to the margins of safety for the fuel oil storage tanks and delivery piping that are required by the design specifications.

The staff noted that AMR items 20m (piping) and 33c (fuel oil tanks) in LRA Table 3.3.2-20 identify that the external surfaces of the fuel oil storage tanks and their associated piping are exposed to a buried soil environment. The staff also noted that, rather than assess whether the original corrosion allowance analysis would remain valid or be acceptable as projected through the expiration of the period of extended operation, the applicant has opted to use 10 CFR 54.21(c)(1)(iii) as the basis for accepting this TLAA and to credit the Buried Piping and Tanks Inspection Program as the AMP for managing loss of material and/or cracking due to corrosion

4-43

in the tank's wall thickness. The staff informed the applicant that the Buried Piping and Tanks Inspection Program was not, by itself, a valid AMP to credit because the AMP did not necessarily guarantee that the external surfaces of the fuel oil storage tanks would be inspected either by an opportunistic bare metal examination or by a supplemental bare metal examination that would be scheduled for the period of extended operation. The staff also informed the applicant that it would either need to: (1) demonstrate that the existing corrosion allowance analysis remains valid for the period of extended operation in accordance with the requirement of 10 CFR 54.21(c)(1)(i), or (2) demonstrate that the corrosion allowance analysis has been projected to through the expiration of the period of extended operation in accordance with the requirement of 10 CFR 54.21(c)(1)(ii), or else (3) demonstrate that the aging effect's impact on the intended function of the fuel oil storage tanks would be managed during the period of extended operation in accordance with the requirement of 10 CFR 54.21(c)(1)(iii). The staff asked the applicant to provide its basis why the Buried Piping and Tanks Inspection Program could be credited in accordance with 10 CFR 54.21(c)(1)(iii) if the program could not ensure that the outside surfaces of the fuel oil storage tanks would be excavated and inspected for evidence of corrosion.

The applicant provided its response to the staff question in a letter dated February 2, 2008. In its response, the applicant stated that, instead or crediting the Buried Piping and Tanks Inspection Program for aging management of corrosion in the diesel fuel oil storage tank, the applicant would amend the application for state that the corrosion allowance analysis for the diesel fuel oil storage tank has been projected to and has been found acceptable in accordance with the staff's requirement in 10 CFR 54.21(c)(1)(ii).

The applicant also stated that, as a conservative measure, the applicant would also continue to credit its Buried Piping and Tanks Inspection Program to manage the impacts of loss of material due galvanic corrosion on the structural integrity of the fuel oil storage tanks. The applicant also stated that LRA Section 4.7.2 and A.3.6.2 would be amended accordingly.

The staff confirmed that, in the applicant's letter of March 20, 2008, the applicant amended the LRA to provide its basis for accepting this TLAA in accordance with 10 CFR 54.21(c)(1)(ii). In this letter, the applicant indicated that the depth of corrosion penetration into the corrosion allowance for the tanks is projected to increase from a 40-year value of 25% penetration to a 60-year value of 51% penetration. The staff noted that the applicant's projected 60-year penetration value for the tanks (51%) is more conservative than the staff's independent 60-year projection of 37.6% penetration into the corrosion allowance for the tanks and meets the acceptance criterion of 100% full penetration into the corrosion allowance for the tanks. Based on this review the staff finds that the applicant has provided an acceptable basis that demonstrates that the projected depth of corrosion penetration into the tanks would be acceptable for the period of extended operation and that the corrosion allowance analysis for the tanks is acceptable in accordance with requirements of 10 CFR 54.21(c)(1)(ii). The applicant's letter of March 20, 2008 also indicated that, as a conservative measure, the applicant would continue to credit its Buried Piping and Tanks Inspection Program to manage the aging effect of loss of material in the diesel fuel oil storage tanks.

The staff also confirmed that, in the applicant's letter of March 20, 2008, the applicant also provided the updated depth of penetration analysis for the diesel fuel storage tank delivery piping. In this analysis, the applicant indicated that the depth of corrosion penetration into the corrosion allowance for the piping is projected to increase from a 40-year value of 50% penetration to a 60-year value of 76% penetration. The staff noted that the applicant's projected 60-year penetration value (76%) for the piping is consistent with the staff's independent 60-year projection of 75.5% penetration into the corrosion allowance for the piping and meets the acceptance criterion of 100% full penetration into the corrosion allowance for the piping. Based on this review, the staff finds that the applicant has provided an acceptable basis that demonstrates that the projected depth of corrosion penetration into the diesel fuel oil tank delivery piping would be acceptable for the period of extended operation and that the corrosion allowance analysis for the piping is acceptable in accordance with requirements of 10 CFR 54.21(c)(1)(ii).

Based on this review, the staff concludes that the applicant has provided an acceptable basis for accepting the corrosion analyses for the diesel fuel oil storage tanks and diesel fuel oil delivery piping in accordance with the requirements of 10 CFR 54.21(c)(1)(ii)

4.7.2.3 UFSAR Supplement

The applicant provided a UFSAR supplement summary description of its TLAA evaluation of fuel oil storage tank corrosion allowance in LRA Section A.3.6.2. The staff noted that the applicant has opted to manage the effects of corrosion on the intended function of the fuel oil storage tanks and associated fuel oil piping using both the Buried Piping and Tanks Inspection Program and the One-Time Inspection Program. The staff verified that, in the applicant's letter of March 20, 2008, the applicant amended the LRA to provide its basis for accepting the corrosion allowance analyses for the diesel fuel oil storage tanks and diesel fuel oil delivery piping in accordance with 10 CFR 54.21(c)(1)(ii), and updated LRA UFSAR Supplement section A.3.6.2 to indicate the following:

> The VEGP Diesel Fuel Oil Storage Tanks and associated piping are not provided with cathodic protection; therefore, a liberal corrosion allowance was included. A calculation performed to evaluate the corrosion allowance included a 40-year assumption and has been determined to be a TLAA.

> The calculation determined the depth of penetration for a hole of approximately 1/32" diameter (0.001 in2) in the coating. The calculation was reviewed for license renewal, and it was determined that depth of penetration due to corrosion would not exceed the corrosion allowance during a 60-year operating life.

> Specifically, consideration of 60 years instead of 40 years in the calculation increases the depth of penetration due to corrosion from 25% to 51% of the corrosion allowance for the tanks and from 50% to 76% of the corrosion allowance for the pipes. Therefore, demonstration is in accordance with 10 CFR 54.21 (c)(1)(ii).

On the basis of its review of the UFSAR supplement, the staff concludes that the summary description of the applicant's actions to address fuel oil storage tank corrosion allowance is adequate because the applicant has updated the UFSAR supplement summary description to provide the applicant's basis for accepting the diesel fuel storage tank and delivery piping corrosion analyses in accordance with the requirements of 10 CFR 54.21(c)(1)(ii).

4.7.2.4 Conclusion

On the basis of its review, as discussed above, the staff concludes that the applicant has demonstrated, pursuant to 10 CFR 54.21(c)(1)(ii), that, for fuel oil storage tank corrosion allowance, the corrosion analyses for the diesel fuel oil storage tanks and diesel fuel oil delivery piping have been projected to the end of period of extended operation. The staff also

4-45

determined that the UFSAR supplement contains an appropriate summary description of the TLAA evaluation, as required by 10 CFR 54.21(d).

4.7.3 Steam Generator Tube, Loss of Material

4.7.3.1 Summary of Technical Information in the Application

LRA Section 4.7.3 summarizes the evaluation of loss of SG tube material for the period of extended operation. UFSAR Section 5.4.2 describes allowances for erosion and corrosion partially based upon a measured rate of loss of material for 40 years. The allowance for erosion and corrosion is much less than the allowed loss of material in the Steam Generator Program. The addition of 50 percent to the loss of material allowances to account for 60 years of operation has no significant effect upon the analysis.

The nominal tube wall thickness is 0.040 inches. The Steam Generator Program – Tube Integrity requires that SG tubes be plugged if they have a 40-percent degradation (0.040*.4 = 0.016 or a wall thickness less than 0.024 inches) from the nominal wall thickness. Results of the specific analysis for allowable tube wall thinning for the Model F SG tubes under normal operating and accident loadings indicate a minimum wall thickness of 0.014 inches to satisfy Regulatory Guide 1.121 stress limits.

The minimum inspection-acceptable wall thickness for new tubes is 0.039 inches. Assumed general wall loss due to corrosion and erosion over 40 years is 0.003 inches, which reduces the tube wall thickness to 0.036 inches. The corrosion rate of 0.003 inches is based on a conservative weight-loss rate for Inconel tubing in flowing 650 °F primary side reactor coolant fluid. The weight loss, when equated to a thinning rate and projected over a 40-year design objective with appropriate reduction after initial hours, is equivalent to 0.000083 inches thinning. The assumed corrosion rate of 0.003 inches allows a conservative 0.002917 inches for general corrosion thinning on the secondary side. An assumed corrosion rate increased by 50 percent from 0.003 to 0.0045 inches has no effect on tube plugging criteria; therefore, the applicant has demonstrated in accordance with 10 CFR 54.21(c)(1)(iii) that the Steam Generator Program – Tube Integrity will manage the loss of material aging effect considered by this TLAA during the period of extended operation.

4.7.3.2 Staff Evaluation

The staff reviewed License Renewal Application Section 4.7.3, to verify, pursuant to Title 10 of the Code of Federal Regulations (10 CFR) 54.21(c)(1)(iii), that the effects of aging on the SG tube integrity will be adequately managed for the period of extended operation.

The VEGP Updated Final Safety Analysis Report (UFSAR) Section 5.4.2 describes allowances for erosion and corrosion that are partially based upon a measured material loss rate for 40 years. The UFSAR allowance for erosion and corrosion is much less than the allowed loss of material in the Vogtle SG Program. Therefore, increasing the projected material loss by 50 percent, to account for 60 years of operation, has been determined to have no significant effect upon the analysis.

The results of tube wall thinning analyses for the Vogtle Model F SG tubes, under normal operating and accident loading conditions, show that a minimum wall thickness of 0.014 inches is necessary to satisfy the stress limits of Regulatory Guide 1.121.

The nominal tube wall thickness of the Vogtle Model F SGs is 0.040 inches. Section 5.5.9, "Steam Generator (SG) Program," of the Vogtle TSs requires that SG tubes be plugged if they become 40 percent degraded from the nominal wall thickness (i.e., a wall thickness less than 0.024 inches). The minimum wall thickness for new tubes is 0.039 inches, which means an acceptable tube has a minimum wall thickness margin of 0.015 inches before plugging is required (0.039 − 0.024 = 0.015).

The projected total loss of tube wall thickness, due to corrosion and erosion on both primary and secondary sides of the tubes over 40 years, is 0.003 inches. The 0.003 inches is based on a conservative weight-loss rate for Inconel tubing in flowing, 650 °F, primary-side reactor coolant. The weight loss, when equated to a thinning rate and projected over a 40-year design objective (with appropriate reduction after initial hours) is equivalent to 0.000083 inches of wall thinning on the primary side of the SG tubes, thereby leaving a conservative 0.002917 inches for general corrosion thinning on the secondary side. Increasing the assumed total loss of tube wall thickness, from 0.003 inches over 40 years, to 0.0045 inches over 60 years, reduces the tube-wall thickness margin (above the tube plugging criteria) from 0.012 inches to 0.0105 inches, but has no effect on the tube plugging criteria. In addition to maintaining a significant margin above the minimum required wall thickness, the licensee performs periodic SG tube inspections to ensure that adequate structural and leakage integrity are maintained.

4.7.3.3 UFSAR Supplement

The applicant provided a UFSAR supplement summary description of its TLAA evaluation of SG tube loss of material in LRA Section A.3.6.3. On the basis of its review of the UFSAR supplement, the staff concludes that the summary description of the applicant's SG tube loss of material program is adequate.

4.7.3.4 Conclusion

Based on its review, the staff concludes that the applicant has demonstrated, pursuant to 10 CFR 54.21(c)(iii), that the loss-of-material aging effect considered by this time-limited aging analysis (TLAA), will be adequately managed during the period of extended operation by the applicant's SG Program under section 5.5.9 of the TS. The staff also determined that the UFSAR supplement contains an appropriate summary description of the TLAA evaluation, as required by 10 CFR 54.21(d).

4.7.4 Cold Overpressure Protection System

4.7.4.1 Summary of Technical Information in the Application

The cold-overpressure mitigation system is described in VEGP UFSAR Section 5.2.2.10. When the RCS is at temperatures below approximately 350°F, it is opened to the Residual Heat Removal (RHR) System for the purposes of removing residual heat from the core, providing a path for letdown to the purification subsystem, and controlling the RCS pressure when the plant is operating in a water solid mode. The RHR System is provided with self-actuated water relief valves to prevent overpressure in this relatively low design pressure system caused either within the system itself or from transients transmitted from the RCS. The RHRS relief valves mitigate pressure transients originating in the RCS to maximum pressure values determined by the relief valve set pressure.

The low design pressure RHR System is normally isolated from the high design pressure RCS during reactor power operation at temperatures above approximately 350°F by two isolation valves in series. Therefore, the RHR System can be inadvertently isolated from the RCS by these same isolation valves. The pressurizer power-operated relief valves (PORVs) and associated logic provide overpressure mitigation for those transients which could occur, if the RHR System isolation valves were inadvertently closed. The PORV logic is manually armed at the system setpoint.

Two pressurizer PORVs are each supplied with actuation logic. The logic for each PORV continuously monitors RCS temperature and pressure, converts an auctioneered RCS temperature to the Appendix G allowable pressure, and then compares the allowable pressure to the actual RCS pressure. As the actual RCS pressure approaches the allowable pressure, a main control board alarm is annunciated. If the RCS pressure continues to increase, an actuation signal is transmitted to a PORV and the valve opens to mitigate the transient. The setpoints for the pressurizer PORVs and arming temperature are developed in conjunction with the P-T curves using the NRC-approved methodology specified in Technical Specification 5.6.6 (Ref. 32) and are specified in the Pressure-Temperature Limits Report (PTLR). The current limits for these cold overpressure protection systems (COPS) setpoints are based on the 36 EFPY steady-state curves (in the PTLR), which are beltline conditions and are not compensated for pressure differences between the pressurizer transmitter and the reactor midplane/beltline or for instrument inaccuracies. These setpoints include an allowance for the 50°F thermal transport effect for heat injection transients. A calculation has been performed to confirm that the setpoints will maintain the system pressure within the established limits when the pressure difference between the pressure transmitter and reactor midplane and maximum temperature/pressure instrument uncertainties are applied to the setpoints. That calculation is a TLAA.

Each time the P-T limit curves are revised (see Section 4.2.5), the impact on the COPS setpoints is evaluated to ensure the functional requirements can be met. The P-T curves have been projected through the period of extended operation; however, operational needs will not require use of these curves before the period of extended operation. As described in the PTLR, the Reactor Vessel Surveillance Program (Appendix B.3.25) updates the P-T limit curves considering the data gained from capsules SNC pulls, and the content and update of the PTLR is in accordance with the VEGP Technical Specification 5.6.6. The P-T limit curves and the associated COPS setpoints will continue to be updated as operational needs dictate to bound the current level of neutron embrittlement (i.e., EFPY) for the unit. Therefore, this TLAA demonstration is made in accordance with 10 CFR 54.21(c)(1)(ii) and (iii).

4.7.4.2 Staff Evaluation

Paragraph IV.A.2 of 10 CFR Part 50, Appendix G, provides the criteria for generating the P-T limits that are required for commercial U.S. light-water reactors. As required by 10 CFR 50.36, licensees owning nuclear power production facilities must include the P-T limits and LTOP setpoints among the limiting conditions for operation (LCOs) in the plant TS. However, on August 31, 1996, the staff issued Generic Letter (GL) 96-03, "Relocation of the Pressure Temperature Limit Curves and Low Temperature Overpressure Protection System Limits," and provided an acceptable process for revising the LCOs for plant-specific P-T limits and LTOP setpoints for relocating the actual P-T limit curves and LTOP setpoints out of the LCOs and into a PTLR. The GL also required that the PTLR be administratively controlled by the Administrative Controls Section of the TS.

The setpoints for the pressurizer PORVs and arming temperature are developed in conjunction with the P-T curves using the NRC-approved methodology specified in Technical Specification 5.6.6, "Technical Specifications Vogtle Electric Generating Plant (VEGP) Unit Nos. 1 and 2 Docket Nos.50-424 and 50-425 Appendix A to License Nos. NPF-68 and NPF-81, December 12, 2006 and are specified in the Pressure-Temperature Limits Report (PTLR) which was approved by the staff in a letter dated January 24, 2005 (ML050180168). The staff's approval of the PTLR permits the applicant to generate the P-T limit curves and the cold overpressure protection systems (COPS) setpoints for the period of extended operation for VEGP, Units 1 and 2, without the need for a license amendment for the curves. The current limits for the COPS setpoints are based on the 36 EFPY steady-state curves (in the PTLR). Since the staff approved the PTLR, the P-T limit curves and the associated COPS setpoints will continue to be updated as operational needs dictate to bound the current level of neutron embrittlement (i.e., EFPY) for the VEGP units. Therefore, the staff finds the applicant's statement to manage COPS set points acceptable because the change in COPS set points will be implemented in accordance with the facility's current licensing basis which meets the regulatory requirements of 10 CFR Part 50, Appendix G.

4.7.4.3 UFSAR Supplement

The applicant provided a UFSAR supplement summary description of its TLAA evaluation of the unit's LTOP limits in LRA Section A.3.6.4. On the basis of its review of the UFSAR supplement, the staff concludes that the summary description of the applicant's actions to address the LTOP limits analysis is adequate.

4.7.4.4 Conclusion

On the basis of its review, as discussed above, the staff concludes that the applicant has demonstrated, pursuant to 10 CFR 54.21(c)(1)(ii), that the analysis has been projected to the end of the period of extended operation, that the effects of aging on the intended function(s) will be adequately managed for the period of extended operation.
The staff also concludes that the UFSAR supplement contains an appropriate summary description of the TLAA evaluation, as required by 10 CFR 54.21(d).

4.7.5 Underclad Cracking of the Reactor Pressure Vessel

4.7.5.1 Summary of Technical Information in the Application

In the SNC letter of March 20, 2008, the applicant amended the LRA with LRA Section 4.7.5, "Underclad Cracking of the Reactor Pressure Vessel," which provides the applicant's time-limited aging analysis for managing underclad cracking in those reactor pressure vessel (RPV) components that are fabricated from SA 508, Class 2 forgings whose internal cladding was welded using a high heat submerged arc weld process. In the SNC letter of February 8, 2008, the applicant identifies that this TLAA is applicable to the following components:

- closure head dome flanges (LRA Table 3.1.2-1, AMR Item 4)
- primary inlet nozzles (LRA Table 3.1.2-1, AMR Item 17),
- primary outlet nozzles (LRA Table 3.1.2-1, AMR Item 20)
- RPV flanges (LRA Table 3.1.2-1, AMR Item 25)

The applicant identifies that the generic underclad cracking analysis in Westinghouse Report No. WCAP-15338 for Westinghouse-designed reactors is applicable to the design of the Vogtle reactors and is a TLAA for these components. In this section of the LRA, the applicant stated that the thermal fatigue crack growth analysis in this report is bounding for 60 years of licensed operation in accordance with 10 CFR 54.21(c)(1)(i).

4.7.5.2 Staff Evaluation

10 CFR 54.21(c)(1) allows for TLAAs to be accepted in accordance with one of the following three options for TLAA acceptance:

(i) the analysis remains acceptable for the period of extended operation

(ii) the analysis has been projected through the expiration of the period of extended operation

(iii) the effects of aging on the intended function(s) will be managed for the period of extended operation

Non-proprietary Westinghouse Report No. WCAP-15338 provides a fracture toughness and flaw growth analysis for underclad cracks that are postulated in the internal cladding of SA-508 Class 2 or 3 alloy steel components in Westinghouse-design RPVs. The flaw growth analysis in the WCAP is based on ASME Section XI, Appendix A, fatigue flaw growth methods and is a generic TLAA for those Westinghouse reactors that credit the report to manage underclad cracking in their SA-508, Class 2 or 3 RVP forging components. The staff accepted the fracture toughness and flaw growth analyses in WCAP-15338 in a safety evaluation (SE) to the Westinghouse Owners Group (WOG) dated October 15, 2001.

The VEGP reactor units are both 4-loop Westinghouse Electric designed PWRs. In its generic safety analysis in WCAP-15338, Westinghouse evaluated the impact of 60 years of operation on underclad growth of cracks that had initiated in the internal cladding of Westinghouse-designed RPV components made from SA-508 Class 2 or 3 alloy steel forged materials. In the staff's SE on WCAP-15338, the staff identified two Renewal Applicant Action Items for PWR applicants that reference WCAP-15338 as a TLAA for their Westinghouse-design LRAs. In the first Renewal Action Item, for applicants owning Westinghouse 2-loop and 4-loop designed PWRs, the staff requested that the renewal applicant should demonstrate that the transients for normal, upset, emergency, faulted, and PTS conditions assessed in WCAP-15338 are bounding for the plant-specific transients for these conditions; otherwise, the staff requested that the applicant perform similar Section XI flaw evaluations using their plant-specific transients to demonstrate that their RPVs with underclad cracks are acceptable though 60 years of licensed operations.

The staff noted that the new TLAA section did not address why the transients for normal, upset, emergency, faulted, and PTS conditions, as assessed in WCAP-15338, were bounding for the VEGP-specific transients that are analyzed for in the normal, upset, emergency, faulted, and PTS condition assessments for the facility. In RAI 4.7.5-1, the staff requested that SNC provide clarification on how the normal, upset, emergency, faulted, and PTS conditions used in WCAP-15338 flaw evaluation bounds the plant-specific transients for these conditions at Vogtle reactor units.

The applicant responded to RAI 4.7.5-1, in SNC Letter No. NL-08-0837, dated May 29, 2008. In its response, the applicant stated that the entire set transient cycles for operations or operational transients associated with normal, upset, emergency, faulted, and PTS conditions are described in LRA Table 4.3.1-1, and that the transients assumed in the fatigue flaw growth analysis of WCAP-15338 over 60 years of licensed operations bound the 60-year transient projections described in the LRA Table 4.3.1-1.

The staff noted that, in Section 5.4 of WCAP-15338, Westinghouse evaluated the fatigue induced crack growth that would occur in postulated flaws that have 2:1, 6:1, and 100:1 length to depth aspect ratios. The staff also noted that in the flaw growth analysis, Westinghouse considered the entire set of design basis transients for Westinghouse designed plants, considering the impact of each design basis transient of the postulated flaw sizes in the analysis. The staff verified that, in Westinghouse's flaw growth process, Westinghouse calculated the crack growth associated with limiting number cycles for each transient in the Westinghouse design basis transient set over a 60-year operational period, adding the crack growth increment to the original postulated flaw size, and then repeating the process until all transient cycles have been accounted for in the final analyzed flaw size.

The staff also verified that the design basis transients for VEGP Units 1 and 2 are described in LRA Table 4.3.1-1. Since the Westinghouse analysis covers the entire set of design basis transients for Westinghouse-designed nuclear reactors, the staff finds that the applicant has provided a valid basis on how the Westinghouse fatigue flaw growth analysis bounds the set of design basis transients for VEGP Units 1 and 2 through 60-years of licensed operations. Based on this review, the staff finds that the applicant has provided an acceptable basis for concluding that 60-year flaw growth analysis in WCAP-15338 is bounding for VEGP Units 1 and 2. Based on this review, the staff finds that the applicant has provided an acceptable response to RAI 4.7.5-1 and the staff's first renewal applicant action item on WCAP-15338 is resolved.

In the second Renewal Action Item identified in the staff's SE on WCAP-15338, the staff requested that the license renewal applicants referencing WCAP-15338 for RPV components provide a summary description of the TLAA evaluation in the UFSAR supplement. The staff's evaluated the applicant's UFSAR supplement for this TLAA in the next section.

The staff finds the applicant's TLAA evaluation of underclad cracking is acceptable in accordance with the requirements of 10 CFR 54.21(c)(1)(i) because the staff has verified that the Westinghouse flaw growth analysis for underclad cracks in Westinghouse RPV components made SA-508 Class 2 or 3 forged materials is bounding for VEGP RPV components that are made from SA-508 Class 2 or 3 materials.

4.7.5.3 UFSAR Supplement

Section 4.7 of the SRP-LR does not currently include a recommended UFSAR supplement summary description for TLAAs on underclad cracking of SA-508, Class 2 or 3 forging components.

In the second Renewal Action Item identified in the staff's SE on WCAP-15338, the staff recommended that the license renewal applicants referencing WCAP-15338 for RPV components provide a summary description of the TLAA evaluation in the UFSAR supplement for their applications. The staff has verified that the applicant has provided the following UFSAR supplement for TLAA Section 4.7.5 in SNC Letter No. NL-08-0195, dated March 20, 2008:

Underclad Cracking of the Reactor Pressure Vessel There is no plant specific evaluation of underclad cracking at VEGP and no such cracks have been identified.

Freedom from underclad cracking is ensured by special evaluation of the procedure qualification for cladding applied on low-alloy steel (SA-508, Class 2) in accordance with Reg. Guide 1.43. However, SNC conservatively includes underclad cracking as a TLAA. Analyses performed by Westinghouse in WCAP-1 5338 demonstrate that growth of under-clad cracks in Westinghouse reactor pressure vessels (RPVs) does not represent a significant challenge to reactor vessel integrity for an operating term of 60 years. The assumptions used as inputs to WCAP-1 5338 are applicable to VEGP. The results of these analyses demonstrate that under-clad cracking of reactor vessel components is not an aging effect requiring management for VEGP. TLAA disposition is in accordance with 10 CFR 54.21 (c)(1)(i)."

In its response to RAI 4.7.5-1, the applicant provided an acceptable technical clarification on how the how the normal, upset, emergency, faulted, and PTS conditions used in WCAP-15338 flaw evaluation bounds the plant-specific transients for these conditions at Vogtle reactor units. The staff has provided its basis for resolving the request raised in RAI 4.7.5-1 in the *Staff Evaluation* section of the SER input for this TLAA. Based on this review, the staff finds the applicant's UFSAR supplement for the TLAA on underclad cracking to be acceptable in accordance with the requirements of 10 CFR 54.21(c)(1)(d) because the UFSAR supplement identifies a description on how the flaw growth analysis in WCAP-15338 is generically applicable and bounding for underclad cracking that are postulated in the VEGP RPV closure head dome flanges, RPV primary inlet nozzles, RPV primary outlet nozzles, and RPV flanges that are fabricated from SA-508, Class 2 alloy steel, and because the applicant's response to RAI 4.7.5-1 provides an acceptable basis why the normal, upset, emergency, faulted, and PTS conditions used in WCAP-15338 are bounding for the Vogtle RPVs and can be used to meet the staff's first renewal applicant action item that is identified in Section 4.1 of the staff's SE on WCAP-15338.

4.7.5.4 Conclusion

On the basis of its review, as discussed above, the staff concludes that the applicant has demonstrated, pursuant to 10 CFR 54.21(c)(1)(iii), that, for fuel oil storage tank corrosion allowance, the effects of aging on the intended function(s) will be adequately managed for the period of extended operation. The staff also determined that the UFSAR supplement contains an appropriate summary description of the TLAA evaluation, as required by 10 CFR 54.21(d).

4.8 Conclusion for TLAAs

The staff reviewed the information in LRA Section 4, "Time-Limited Aging Analyses." On the basis of its review, the staff concludes, that the applicant has provided a sufficient list of TLAAs, as defined in 10 CFR 54.3 and that the applicant has demonstrated that: (1) the TLAAs will remain valid for the period of extended operation, as required by 10 CFR 54.21(c)(1)(i); (2) the TLAAs have been projected to the end of the period of extended operation, as required by 10 CFR 54.21(c)(1)(ii); or (3) that the effects of aging on intended function(s) will be adequately managed for the period of extended operation, as required by 10 CFR 54.21(c)(1)(iii). The staff also reviewed the UFSAR supplement for the TLAAs and finds that the supplement contains descriptions of the TLAAs sufficient to satisfy the requirements of 10 CFR 54.21(d).

With regard to these matters, the staff concludes that there is reasonable assurance that the activities authorized by the renewed licenses will continue to be conducted in accordance with the CLB, and that any changes made to the CLB, in order to comply with 10 CFR 54.29(a), are in accordance with the Atomic Energy Act of 1954, as amended, and NRC regulations.

THIS PAGE INTENTIONALLY LEFT BLANK.

SECTION 5

REVIEW BY THE ADVISORY COMMITTEE ON REACTOR SAFEGUARDS

The NRC staff issued its safety evaluation report related to the renewal of operating licenses for Vogtle Electric Generating Plant (VEGP), Units 1 and 2 on September 12, 2008. On November 05, 2008, the applicant presented its license renewal application, and the staff presented its review findings to the ACRS Plant License Renewal Subcommittee. The Staff reviewed the applicant's comments on the SER and completed it review of the license renewal application. The staff's evaluation is documented in an SER that was issued by letter dated March 13, 2009.

During the 561st meeting of the ACRS, April 2, 2009 through April 4, 2009, the ACRS completed its review of the VEGP license renewal application and the NRC staff's SER. The ACRS documented its findings in a letter to the Commission dated April 17, 2009. A copy of this letter is provided on the following pages of this SER Section.

UNITED STATES
NUCLEAR REGULATORY COMMISSION
ADVISORY COMMITTEE ON REACTOR SAFEGUARDS
WASHINGTON, DC 20555 - 0001

April 10, 2009

The Honorable Dale E. Klein
Chairman
U. S. Nuclear Regulatory Commission
Washington, DC 20555-0001

SUBJECT: REPORT ON THE SAFETY ASPECTS OF THE LICENSE RENEWAL
             APPLICATION FOR THE VOGTLE ELECTRIC GENERATING PLANT,
             UNITS 1 AND 2

Dear Chairman Klein:

During the 561st meeting of the Advisory Committee on Reactor Safeguards, April 2-4, 2009, we
completed our review of the license renewal application for the Vogtle Electric Generating Plant
(VEGP), Units 1 and 2, and the final Safety Evaluation Report (SER) prepared by the NRC staff.
Our Plant License Renewal Subcommittee also reviewed this matter during its meeting on
November 5, 2008. During these reviews, we had the benefit of discussions with
representatives of the NRC staff and the applicant, Southern Nuclear Operating Company Inc.
(SNC). We also had the benefit of the documents referenced. This report fulfills the
requirement of 10 CFR 54.25 that the ACRS review and report on all license renewal
applications.

CONCLUSION AND RECOMMENDATION

1.   The programs established and committed to by the applicant to manage age-related
         degradation provide reasonable assurance that VEGP, Units 1 and 2 can be operated in
         accordance with its current licensing basis for the period of extended operation without
         undue risk to the health and safety of the public.

2.   The SNC application for renewal of the operating licenses of VEGP Units 1 and 2 should
         be approved.

BACKGROUND AND DISCUSSION

VEGP consists of two Westinghouse 4-loop pressurized water reactors with large dry ambient
containments and is located approximately 26 miles southeast of Augusta, Georgia. The
current licensed power rating of each of the VEGP units is 3,625 megawatts thermal with a
gross electrical output of approximately 1,250 megawatts. SNC requested renewal of the VEGP
operating licenses for 20 years beyond the current license terms, which expire on
January 16, 2027, for Unit 1 and February 9, 2029, for Unit 2.

In the final SER, the staff documented its review of the license renewal application and other
information submitted by the applicant or obtained from the staff audit and inspection at the

plant site. The staff reviewed the completeness of the applicant's identification of the structures, systems, and components (SCCs) that are within the scope of license renewal; the integrated plant assessment process; the applicant's identification of the plausible aging mechanisms associated with passive, long-lived components; the adequacy of the applicants Aging Management Programs (AMPs); and the identification and assessment of time-limited aging analyses (TLAAs) requiring review.

In the VEGP license renewal application, SNC identified the SSCs that fall within the scope of license renewal. For these SSCs, the applicant performed a comprehensive aging management review. Based on this review, the applicant will implement 38 AMPs for license renewal, including 24 existing programs of which 10 are enhanced and 10 have exceptions, and 14 new programs.

The VEGP application either demonstrates consistency with the Generic Aging Lessons Learned (GALL) Report or documents deviations to the specified approaches in this Report. The VEGP application includes several exceptions to the GALL Report. We reviewed these exceptions and agree with the staff that they are acceptable. The staff conducted a license renewal audit and inspection at VEGP. The audit verified the appropriateness of the scoping and screening methodology, AMPs, aging management review, and TLAAs. The site inspection verified that the license renewal requirements are appropriately implemented. Based on the audit and inspection, the staff concluded in the final SER that the proposed activities will reasonably manage the effects of aging of SSCs identified in the application and that the intended functions of these SSCs will be maintained during the period of extended operation. We agree with this conclusion.

The applicant identified the systems and components requiring TLAAs and reevaluated them for the period of extended operation. The staff concluded that the applicant has provided an adequate list of TLAAs. Further, the staff concluded that the applicant has met the requirements of the License Renewal Rule by demonstrating that the TLAAs will remain valid for the period of extended operation, or that the TLAAs have been projected to the end of the period of extended operation, or that the aging effects will be adequately managed for the period of extended operation.

During its site inspection, the staff observed water in manholes which contain medium voltage cables that are important to safety. We did not see any evidence of environmental qualification of these cables and associated splices for submerged operation. The staff has identified water in manholes as a generic, current operating plant issue in Information Notice 2002-12, "Submerged Safety-Related Electrical Cables," and Generic Letter 2007-01, "Inaccessible or Underground Power Cable Failures that Disable Accident Mitigation Systems or Cause Plant Transients." During the current period of operation, the staff will address the issue of water in manholes through the Reactor Oversight Process.

We agree with the staff that there are no issues related to the matters described in 10 CFR 54.29(a)(1) and (a)(2) that preclude renewal of the operating licenses for VEGP, Units 1 and 2. The programs established and committed to by SNC provide reasonable assurance that the VEGP can be operated in accordance with its current licensing basis for the period of extended operation without undue risk to the health and safety of the public.

The SNC application for renewal of the operating licenses for VEGP, Units 1 and 2 should be approved.

Dr. Said Abdel-Khalik did not participate in the Committee's deliberations regarding this matter.

Sincerely,

/RA/

Mario V. Bonaca
Chairman

References:

1.   U.S. Nuclear Regulatory Commission, Safety Evaluation Report Related to the License Renewal of Vogtle Electric Generating Plant Units 1 and 2, dated March 2009 (ML090710010)

2.   Letter from L. M. Stinson, Southern Nuclear Operating Company, Incorporated, to U.S. Nuclear Regulatory Commission, transmitting the Vogtle Electric Generating Plant, Units 1 and 2, Application for License Renewal dated June 27, 2007 (ML071840351)

3.   Letter from K. M. Kennedy, U.S. Nuclear Regulatory Commission, to Tom E. Tynan, Southern Nuclear Operating Company, Incorporated, transmitting Vogtle Electric Generating Station, Units 1 and 2; NRC Inspection Report 05000424-08-006 and 05000425-08-006 dated July 18, 2008 (ML082000869)

4.   Letter from D. J. Ashley, U.S. Nuclear Regulatory Commission, to Tom E. Tynan, Southern Nuclear Operating Co, Incorporated, transmitting the Audit Summary Report Regarding the License Renewal Application for the Vogtle Electric Generating Plant Units 1 and 2 dated October 1, 2008 (ML082680295)

5.   U.S. Nuclear Regulatory Commission, NUREG-1801, Volumes 1 & 2, Revision 1, "Generic Aging Lessons Learned Report," dated September 2005 (ML052700171)

6.   U. S. Nuclear Regulatory Commission, Information Notice 2002-12, "Submerged Safety-related Electrical Cables," dated March 21, 2002 (ML020790238)

7.   U. S. Nuclear Regulatory Commission, Generic Letter 2007-01, "Inaccessible or Underground Power Cable Failures that Disable Accident Mitigation Systems or Cause Plant Transients," dated February 7, 2007 (ML070360665)

SECTION 6
CONCLUSION

The staff of the United States Nuclear Regulatory Commission (NRC or the staff) reviewed the license renewal application (LRA) for Vogtle Electric Generating Plant, Units 1 and 2, in accordance with NRC regulations and NUREG-1800, Revision 1, "Standard Review Plan for Review of License Renewal Applications for Nuclear Power Plants," dated September 2005. Title 10, Section 54.29, of the *Code of Federal Regulations* (10 CFR 54.29) sets the standards for issuance of a renewed license.

On the basis of its review of the LRA, the staff concludes that the requirements of 10 CFR 54.29(a) have been met.

The staff noted that any requirements of 10 CFR Part 51, Subpart A, are documented in NUREG-1437, "Generic Environmental Impact Statement for License Renewal of Nuclear Plants (GEIS)," Supplement 34, "Regarding Vogtle Electric Generating Plant, Units 1 and 2 Draft Report for Comment," dated April 2008.

APPENDIX A

VEGP UNITS 1 AND 2 LICENSE RENEWAL COMMITMENTS

During the review of the Vogtle Electric Generating Plant (VEGP), Units 1 and 2, license renewal application (LRA) by the staff of the United States (US) Nuclear Regulatory Commission (NRC) (the staff), Southern Nuclear Operating Company, Inc. (the applicant) made commitments related to aging management programs (AMPs) to manage aging effects for structures and components. The following table lists these commitments along with the implementation schedules and sources for each commitment.

VEGP License Renewal Future Action Commitment List

ITEM NO.	COMMITMENT	UFSAR SUPPLEMENT LOCATION	SCHEDULE	RELATED LRA SECTION S/ REFS
1	Implement the ACCW System Carbon Steel Components Program as described in VEGP LRA Section B.3.1.	A.2.1	Prior to the period of extended operation	B.3.1
2	Implement the Bolting Integrity Program as described in VEGP LRA Section B.3.2.	A.2.2	Prior to the period of extended operation	B.3.2
3	Enhance Boric Acid Corrosion Control Program documents to address the effects of borated water leakage onto materials other than steels, including electrical components (e.g., electrical connectors), that are susceptible to boric acid corrosion.	A.2.3	Prior to the period of extended operation	B.3.3

A-1

VEGP License Renewal Future Action Commitment List

ITEM NO.	COMMITMENT	UFSAR SUPPLEMENT LOCATION	SCHEDULE	RELATED LRA SECTION S/ REFS
4	Implement the Buried Piping and Tanks Inspection Program as described in VEGP LRA Section B.3.4.	A.2.4	Prior to the period of extended operation	B.3.4
5	Implement the CASS RCS Fitting Evaluation Program as described in VEGP LRA Section B.3.5.	A.2.5	Prior to the period of extended operation	B.3.5
6	Enhance Closed Cooling Water Program documents to indicate the components in each system that are most susceptible to various corrosion mechanisms and to ensure that corrosion monitoring is appropriately accomplished. This qualitative assessment will be based on an understanding of corrosion principles associated with closed cooling water chemistries and on review of system, plant, and industry operating experience. Parameters considered in the review will include system flow parameters (focusing on identification of stagnant regions and on intermittently operated systems), normal operating temperatures, and component geometries (e.g. creviced areas).	A.2.6	Prior to the period of extended operation	B.3.6
7	Implement the External Surfaces Monitoring Program as described in VEGP LRA Section B.3.8.	A.2.8	Prior to the period of extended operation	B.3.8

A-2

VEGP License Renewal Future Action Commitment List

ITEM NO.	COMMITMENT	UFSAR SUPPLEMENT LOCATION	SCHEDULE	RELATED LRA SECTION S/ REFS
8	Implement the following enhancements to the Fire Protection Program:	A.2.9	Prior to the period of extended operation, except for sprinkler head replacement or testing;	B.3.9
	• Wall thickness evaluations will be performed on water suppression piping systems using non-intrusive volumetric testing or visual inspections to ensure that wall thicknesses are within acceptable limits. Initial wall thickness evaluations will be performed before the end of the current operating term. Subsequent evaluations will be performed at plant specific intervals during the period of extended operation. The plant specific inspection intervals will be determined based on previous evaluations and site operating experience.			
	• A sample of sprinkler heads will be inspected using the guidance of NFPA 25 "Inspection, Testing and Maintenance of Water-Based Fire Protection Systems" (1998 Edition), Section 2-3.1.1, or NFPA 25 (2002 Edition), Section 5.3.1.1.1. Where sprinkler heads have been in service for 50 years, they will be replaced or representative samples from one or more sample areas will be submitted to a recognized testing laboratory for field service testing. This sampling will be performed every 10 years after the initial field service testing. The 50 years of time in service begins when the system was placed in service, not when the plant became operational.		Sprinkler head replacement or testing will be implemented prior to 50 years from time system was placed in service.	
	• Fire Protection Program procedures will be revised to provide more detailed instructions for visual inspection of Fire Pump Diesel fuel supply lines for leakage, corrosion, and general degradation while the engine is running during fire suppression system pump tests.			

A-3

VEGP License Renewal Future Action Commitment List

ITEM NO.	COMMITMENT	UFSAR SUPPLEMENT LOCATION	SCHEDULE	RELATED LRA SECTION S/ REFS
9	Enhance the Flux Thimble Tube Inspection Program by preparing an overall program procedure documenting the Flux Thimble Tube Inspection Program administration and implementing activities credited for license renewal.	A.2.11	Prior to the period of extended operation	B.3.11
10	Not Used at VEGP			
11	Implement the following enhancements to the Generic Letter 89-13 Program: • Develop an overall program procedure for the Generic Letter 89-13 Program to describe the various program activities that comprise the Generic Letter 89-13 Program and their implementing controls such as chemistry procedures, maintenance activities, scheduled surveillances, or other mechanisms. • Add inspection of the NSCW Transfer Pumps' casings and bolting. • Add the NSCW Cooling Tower spray nozzles as a specific item to be inspected during the cooling tower inspection.	A.2.12	Prior to the period of extended operation	B.3.12

A-4

VEGP License Renewal Future Action Commitment List

ITEM NO.	COMMITMENT	UFSAR SUPPLEMENT LOCATION	SCHEDULE	RELATED LRA SECTION S/ REFS
12	Implement the Nickel Alloy Management Program for Non-Reactor Vessel Closure Head Penetration Locations as described in VEGP LRA Section B.3.14. The program will be based on the following commitments: (1) SNC will continue to participate in industry initiatives directed at resolving PWSCC issues, such as owner's group programs and the Electric Power Research Institute Materials Reliability Program. This is an ongoing commitment. (2) SNC will comply with applicable NRC Orders. This is an ongoing commitment. (3) SNC will submit a program inspection plan for VEGP that includes implementation of applicable Bulletins, Generic Letters, and staff accepted industry guidance. The inspection plan will also include assessments of each of the ten aging management program elements defined in Section A.1.2.3 of NUREG-1800 Revision 1. The inspection plan will be submitted to the staff for review and approval not less than 24 months prior to entering the period of extended operation for VEGP Units 1 and 2.	A.2.14	Prior to the period of extended operation Numbered items are implemented as noted in each item.	B.3.14
13	The Nickel Alloy Management Program for Reactor Vessel Closure Head Penetrations will implement commitments for reactor vessel closure head penetrations associated with nickel alloys from: (1) NRC Orders, Bulletins, and Generic Letters, and; (2) Staff-accepted industry guidelines.	A.2.15	Ongoing	B.3.15

VEGP License Renewal Future Action Commitment List

ITEM NO.	COMMITMENT	UFSAR SUPPLEMENT LOCATION	SCHEDULE	RELATED LRA SECTION S/ REFS
14	Implement the following enhancements to the Oil Analysis Program: • An overall program procedure or guideline will be prepared to formalize the sampling and analysis activities performed. • Viscosity, relative level of oxidation, and flash point of lubricating oil samples will be determined for components where the lubricating oil is changed based on its analyzed condition (instead of being changed on a regular schedule regardless of condition). The relative level of oxidation of the lubricating oil will be monitored by analysis of the neutralization number or other appropriate parameter(s). Flash point monitoring will be performed for those components which have the potential for contamination of the lubricating oil with a light hydrocarbon such as fuel oil. • For both components with periodic lubricating oil changes and components where the lubricating oil is changed based on analyzed condition, if a lubricating oil sample exceeds the limits established for the wear metal content screening, the lubricating oil from that component will be subjected to additional testing. The additional testing may include detailed particle counting, elemental analysis, or analytical ferrography as necessary to validate the initial screening results and to diagnose the source of the particulates.	A.2.16	Prior to the period of extended operation	B.3.16

VEGP License Renewal Future Action Commitment List

ITEM NO.	COMMITMENT	UFSAR SUPPLEMENT LOCATION	SCHEDULE	RELATED LRA SECTION S/ REFS
15	Implement the One-Time Inspection Program as described in VEGP LRA Section B.3.17.	A.2.17	Inspections will be performed within a window of ten years prior to the period of extended operation.	B.3.17
16	Implement the One-Time Inspection Program for ASME Class 1 Small Bore Piping as described in VEGP LRA Section B.3.18.	A.2.18	Inspections will be performed within a window of ten years prior to the period of extended operation.	B.3.18
17	Implement the One-Time Inspection Program for Selective Leaching as described in VEGP LRA Section B.3.19.	A.2.19	Inspections will be performed within a window of ten years immediately prior to the period of extended operation.	B.3.19

VEGP License Renewal Future Action Commitment List

ITEM NO.	COMMITMENT	UFSAR SUPPLEMENT LOCATION	SCHEDULE	RELATED LRA SECTION S/ REFS
18	Enhance the Periodic Surveillance and Preventive Maintenance Activities to include the following: • Steam Generator Blowdown Secondary Sample Bath Shell inspections • Steam Generator Blowdown Corrosion Product Monitor cooler shell inspections • Potable Water System water heater housing inspections (for the in-scope water heaters)	A.2.21	Prior to the period of extended operation	B.3.21
19	Implement the Piping and Duct Inspection Program as described in LRA Section B.3.22.	A.2.22	Prior to the period of extended operation	B.3.22

VEGP License Renewal Future Action Commitment List

ITEM NO.	COMMITMENT	UFSAR SUPPLEMENT LOCATION	SCHEDULE	RELATED LRA SECTION S/ REFS
20	Implement the Reactor Vessel Internals Program as described in LRA Section B.3.24. The program will be based on the following commitments: (1) SNC will participate in the industry program for investigating and managing aging effects on reactor internals. This is an ongoing commitment. (2) SNC will evaluate and implement the results of the industry programs, such as the Electric Power Research Institute Material Reliability Program, as applicable to the VEGP reactor internals. This commitment will be fully implemented prior to the period of extended operation. (3) SNC will submit an inspection plan for the VEGP reactor internals to the NRC for review and approval not less than 24 months before entering the period of extended operation for VEGP Units 1 and 2. This inspection plan will address the bases, inspection methods, and acceptance criteria associated with aging management of the reactor vessel thermal sleeves and the core support lugs (along with the associated support pads and attachment welds).	A.2.24	Program implementation to be completed prior to the period of extended operation; Numbered items are implemented as noted in each item.	B.3.24

VEGP License Renewal Future Action Commitment List

ITEM NO.	COMMITMENT	UFSAR SUPPLEMENT LOCATION	SCHEDULE	RELATED LRA SECTION S/ REFS
21	Implement the following enhancements to the Reactor Vessel Surveillance Program: (1) Prior to completion of testing of the last surveillance capsule in each unit, action will be taken to ensure that tested and untested specimens from all capsules removed from the VEGP reactor vessels remain in storage. (2) Alternate dosimetry will be installed to monitor neutron fluence on the reactor vessel after removal of the last surveillance capsule in that unit. This enhancement will be implemented prior to removal of the last surveillance capsule in each unit.	A.2.25	As noted in the numbered items	B.3.25
22	Not Used at VEGP			

VEGP License Renewal Future Action Commitment List

ITEM NO.	COMMITMENT	UFSAR SUPPLEMENT LOCATION	SCHEDULE	RELATED LRA SECTION S/ REFS
23	Implement the following enhancements to the Structural Monitoring Program: • The scope of the Structural Monitoring Program will be expanded to include the additional structures that require monitoring for license renewal. • The scope of inspection for structures that require monitoring for license renewal will be clarified. An area-based inspection will be performed unless a detailed inspection scope is provided. • The Structural Monitoring Program scope for hangers and supports will be clarified. • Program requirements will be revised to include periodic ground water monitoring to confirm that groundwater chemistry remains non-aggressive as defined in NUREG 1801. • Underwater inspection of the NSCW cooling tower basins, including appropriate inspection and acceptance criteria, will be added to the Structural Monitoring Program.	A.2.32	Prior to the period of extended operation	B.3.32
24	Enhance the Structural Monitoring Program - Masonry Walls to include monitoring of masonry walls in the structures that are in scope for license renewal, but are not currently monitored under the program.	A.2.33	Prior to the period of extended operation	B.3.33

VEGP License Renewal Future Action Commitment List

ITEM NO.	COMMITMENT	UFSAR SUPPLEMENT LOCATION	SCHEDULE	RELATED LRA SECTION S/ REFS
25	Implement the Non-EQ Cables and Connections Program as described in LRA Section B.3.34.	A.2.34	Implement program and complete first inspection prior to the period of extended operation	B.3.34
26	Implement the Non-EQ Inaccessible Medium-Voltage Cables Program as described in LRA Section B.3.35.	A.2.35	Implement program and complete first inspection prior to the period of extended operation	B.3.35
27	Implement the Non-EQ Cable Connections One-Time Inspection Program as described in LRA Section B.3.36.	A.2.36	Inspections will be performed within a window of five years immediately prior to the period of extended operation.	B.3.36

	VEGP License Renewal Future Action Commitment List			
ITEM NO.	COMMITMENT	UFSAR SUPPLEMENT LOCATION	SCHEDULE	RELATED LRA SECTION S/ REFS
28	Implement the following enhancements to the Fatigue Monitoring Program: • Implementing documents will be revised to address the effect of the full structural weld overlays applied to the pressurizer spray and surge nozzles on the stress-based module calculation of CUF. • The VEGP UFSAR will be revised to require fatigue monitoring of the Accumulator/RHR nozzles and pressurizer heater penetrations. • Implementing documents will be revised to reduce acceptable CUF values to account for environmental fatigue effects for those NUREG-6260 locations monitored for fatigue. • The implementing procedure for the Fatigue Monitoring Program will be enhanced to explicitly require that the corrective action initiated for exceeding the acceptance criteria of a CUF less than or equal to 1.0 includes a review to identify and assess any additional affected reactor coolant pressure boundary locations.	A.2.38	No later than two years prior to the period of extended operation	B.3.38

VEGP License Renewal Future Action Commitment List

ITEM NO.	COMMITMENT	UFSAR SUPPLEMENT LOCATION	SCHEDULE	RELATED LRA SECTION S/ REFS
29	To ensure that the fatigue cycle limits are not exceeded, SNC will replace the steam generator secondary side handhole bolts after 30 years of service. The handhole bolts have been previously replaced and are scheduled to be replaced again during the spring outages in 2026 and 2028 for Units 1 and 2, respectively. Alternatively, a less restrictive replacement schedule may be developed and documented based on potential updated analyses initiated by the Bolting Integrity Program.	A.3.2.5	As stated in the commitment	4.3.5 B.3.2
30	To ensure that the fatigue cycle limits are not exceeded, SNC will replace the steam generator secondary side manway bolts after 30 years of service. The manway bolts have never been replaced and are scheduled for replacement during the spring outages in 2017 and 2019 for Units 1 and 2, respectively. Alternatively, a less restrictive replacement schedule may be developed and documented based on potential updated analyses initiated by the Bolting Integrity Program.	A.3.2.5	As stated in the commitment	4.3.5 B.3.2
31	The VEGP Pressure-Temperature Limits Report (for each unit) will be updated to address neutron embrittlement for a 60-year operating life, including any changes to the cold-overpressure mitigation system setpoints.	A.3.1.5 A.3.6.4	Prior to the period of extended operation	4.2.5 4.7.4

VEGP License Renewal Future Action Commitment List

ITEM NO.	COMMITMENT	UFSAR SUPPLEMENT LOCATION	SCHEDULE	RELATED LRA SECTION S/ REFS
32	Implement a replacement schedule for the small diameter (≤ 2-inch) flexible hoses described below:	None	Prior to the period of extended operation	
	• Radiation Monitoring System flexible hoses associated with the vent stack radiation monitor sample line.			2.3.3.25
	• Emergency Diesel Generator System flexible hoses associated with the fuel oil supply lines from the fuel oil headers to the fuel injector pumps.			2.3.3.20
	• ACCW System flexible hoses associated with the normal charging pump motor coolers.			2.3.3.6
	• Hydrogen Recombiner and Monitoring System flexible hoses associated with the calibration gas and oxygen bottles.			2.3.3.20
	• Main Steam System flexible hoses between the ARV hydraulic actuator and the hand pump unit.			2.3.4.1
	• Drain System flexible hoses installed on the Containment Bldg Tendon Gallery Sump Pump discharge lines.			2.3.3.23
	• Fire Protection System flexible hoses associated with the fire pump diesel fuel oil system.			2.3.3.19
33	Ensure the fatigue monitoring limits implemented as part of the Fatigue Monitoring Program are adequate to ensure that charging and letdown intermediate break location CUF values remain less than 0.1 for 60 years of operation.	A.3.2.1	No later than two years prior to the period of extended operation	4.3.1.7

A-15

VEGP License Renewal Future Action Commitment List

ITEM NO.	COMMITMENT	UFSAR SUPPLEMENT LOCATION	SCHEDULE	RELATED LRA SECTION S/ REFS
34	The cranes within the scope of the Overhead and Refueling Crane Inspection Program are routinely inspected; however the existing procedures do not explicitly identify inspection of structural components for excessive wear, corrosion, and misalignment in all cases. As a result, SNC will enhance applicable plant procedures to explicitly identify inspection of crane rails and crane structural components for loss of material due to corrosion and wear, and for indication of rail misalignment: a) Procedure 93246-C "Polar Crane" will be enhanced to include inspection of crane rails and crane structural components (e.g., bridge) for loss of material due to corrosion; inspection of crane rails for loss of material due to wear, and for indication of rail misalignment. b) Procedure 27315-C "Spent Fuel Cask Crane" will be enhanced to include inspection of crane rails for loss of material due to corrosion. c) Procedure 27340-C "Refueling Machine" will be enhanced to include inspection of crane rails and crane structural components (e.g., bridge) for loss of material due to corrosion, and for indication of rail misalignment. d) Procedure 27342-C "Fuel Handling Machine Bridge Crane" will be enhanced to include inspection of crane rails and crane structural components (e.g., bridge) for loss of material due to corrosion; inspection of crane rails for loss of material due to wear, and for indication of rail misalignment.	A.2.20	Prior to the period of extended operation	B.3.20

A-16

VEGP License Renewal Future Action Commitment List

ITEM NO.	COMMITMENT	UFSAR SUPPLEMENT LOCATION	SCHEDULE	RELATED LRA SECTION S/ REFS
35	SNC will revise the FatiguePro software to calculate a minimum projected value of 1 for any events that may potentially occur,	A.2.28	No later than two years prior to the period of extended operation	A.2.28
36	Once the NRC has accepted a process for addressing PWSCC of Alloy 82/182 welds in LBB evaluations, SNC will verify that the LBB evaluation in WCAP-10551-P, Addendum 1 meets the conditions of that process or have it re-performed using the acceptable process.	A.3.6.1	No later than two years prior to the period of extended operation	A.3.6.1

VEGP License Renewal Future Action Commitment List

ITEM NO.	COMMITMENT	UFSAR SUPPLEMENT LOCATION	SCHEDULE	RELATED LRA SECTION S/ REFS
37	To ensure the Boral spent fuel racks will continue to perform their intended function during the period of extended operation, VEGP will monitor spent fuel pool aluminum concentrations and implement corrective actions if adverse trends are identified. Additionally, SNC will monitor industry experience related to Boral and will take appropriate actions if significant degradation of Boral is identified.	A.2.28	Prior to the period of extended operation	A.2.28
	SNC will also perform a baseline inspection and a follow-up inspection to measure the effectiveness of the Boral neutron absorber panels on Unit 1 to provide reasonable assurance that the panels will continue to perform their reactivity control function during the period of extended operation. These inspections will be included in the One-Time Inspection Program which is to be implemented for license renewal. The baseline inspection will be performed prior to the period of extended operation. The follow-up inspection will be performed at a date to be determined based on the results of the baseline inspection and relevant industry guidance, not to exceed ten years after the baseline inspection.	A.2.17	The baseline inspection will be performed within a window of ten years immediately preceding the period of extended operation. The follow-up inspection will be performed as described in the commitment.	B.3.17

A-18

VEGP License Renewal Future Action Commitment List

ITEM NO.	COMMITMENT	UFSAR SUPPLEMENT LOCATION	SCHEDULE	RELATED LRA SECTION S/ REFS
38	SNC commits to revise the FatiguePro initial CUF values for the Unit 1 and Unit 2 hot leg surge nozzles, pressurizer surge nozzles, and pressurizer heater penetrations to double the current values and recalculate the current and projected CUFs.	A.2.28	No later than two years prior to entering the period of extended operation.	A.2.28
39	SNC commits to implement a fatigue management software program that uses six stress components in the stress based fatigue calculation. The software will be appropriately benchmarked against an ASME NB-3200 fatigue analysis and the stress based fatigue monitoring locations will be modeled with the as-built configuration. The new software will be used to re-project 60-year CUF values for the monitored locations. When those locations were evaluated for environmental effects on fatigue, the new software will also be used to demonstrate that the environmental effects on fatigue will be adequately managed for those locations during the period of extended operation. This software will be put in service at least two years prior to the period of extended operation.	A.2.28	No later than two years prior to entering the period of extended operation.	A.2.28

A-19

VEGP License Renewal Future Action Commitment List

ITEM NO.	COMMITMENT	UFSAR SUPPLEMENT LOCATION	SCHEDULE	RELATED LRA SECTION S/ REFS
40	The following changes will be made to the Structural Monitoring Program in order to enhance evaluation and trending of findings: 1) Guidance will be developed regarding proper documentation of condition adverse to quality and its probable causes for any CR written against a finding during structural monitoring program walkdown. 2) For any finding (e.g., crack, leakage, etc.) guidance will be developed for data to be collected and evaluated. 3) More explicit direction will be given for trending of the problems.	A.2.32	No later than two years prior to the period of extended operation	A.2.32
41	The following changes will be made to the IWE and IWL programs in order to enhance evaluation and trending of findings: 1) More explicit direction will be given to the Registered Professional Engineer for trending and evaluating conditions (including evidence of tendon grease leakage) identified during concrete visual examinations.	A.2.30 A.2.31	No later than two years prior to the period of extended operation	A.2.30 A.2.31
42	SNC commits to perform an inspection of an emergency diesel generator fuel oil day tank vent line. This inspection will determine whether a debris screen is installed on the open end of the vent line. If a screen is installed, the inspection will further determine the material of construction of the debris screen.	A.2.8	Prior to the Period of extended operation	B.3.8

A-20

APPENDIX B

CHRONOLOGY

This appendix lists chronologically the routine licensing correspondence between the staff of the United States (US) Nuclear Regulatory Commission (NRC) (the staff) and Southern Nuclear Operating Company, Inc. (SNC). This appendix also lists other correspondence on the staff's review of the Vogtle Electric Generating Plant (VEGP), Units 1 and 2 license renewal application (LRA) (under Docket Nos. 50-424 and 50-425).

APPENDIX B: CHRONOLOGY	
Date	**Subject**
04/05/2007	Press Release, NRC to Discuss 2006 Performance at Vogtle Nuclear Plant., (ML070950086)
06/27/2007	Letter from Stinson L M, Southern Nuclear Operating Co, Inc to NRC Document Control Desk, Vogtle, Units 1 & 2, Application for License Renewal., (ML071840351)
06/30/2007	Environmental Report from Southern Nuclear Operating Co, Inc to NRC Document Control Desk, Vogtle, Units 1 and 2, Applicant's Environmental Report Operating License Renewal, (ML071840357)
06/30/2007	License-Application for Facility Operating License from Southern Nuclear Operating Co, Inc to NRC Document Control Desk, Vogtle, Units 1 and 2, License Renewal Application, (ML071840360)
07/10/2007	Letter from J. P. Leous, NRC to M. E. Sikes, Maintenance of Reference Materials Related to the Review of the Vogtle Electric Generating Plant Units 1 and 2 License Renewal Application at the Burke County Library., (ML071860391)
07/25/2007	Press Release-07-090 - License Renewal Application for Vogtle Nuclear Plant Available for Public Inspection., (ML072060326)
07/26/2007	Federal Register Notice - Receipt of Vogtle License Renewal Application., (ML071840090)
07/26/2007	Letter from P. T. Kuo to Tom E. Tynan, Southern Nuclear Operating Co, Inc, Receipt of Vogtle License Renewal Application - Acknowledgment Letter (ML071860194)
08/01/2007	Meeting Agenda, from D. J. Ashley to Louise A. Lund, 08/21/2007, Notice of Public Information Meeting to Discuss the License Renewal Process for Vogtle, Units 1 and 2, Application Review, (ML072010412)
08/15/2007	Letter from P. T. Kuo NRC to Tom E. Tynan, Southern Nuclear Operating Co, Inc, Determination of Acceptability and Sufficiency for Docketing, Proposed Review Schedule, and Opportunity for a Hearing Regarding the Application from Southern Nuclear Operating Company, Inc., for Renewal of the Operating Licenses for the Vogtle Electric., (ML072010369)
08/15/2007	Federal Register Notice, : Vogtle Electric Generating Plant, Units 1 and 2., (ML072130084)

APPENDIX B: CHRONOLOGY	
Date	**Subject**
08/16/2007	Press Release-07-103: NRC to Discuss Process for Review of License Renewal Application for Vogtle Nuclear Power Plant August 21., (ML072280386)
08/21/2007	Federal Register Notice, FRN - Notice of Intent to Prepare an Environmental Impact Statement and Conduct Scoping Process., (ML072140337)
08/22/2007	Press Release-07-107: NRC Announces Opportunity to Request Hearing on Application to Renew Operating Licenses for Vogtle Nuclear Power Plant., (ML072340067)
08/23/2007	Letter from Tom E. Tynan, Southern Nuclear Operating Co, Inc to NRC Document Control Desk, Vogtle - Requests that Listed Individuals Receive Correspondence Related to Plant License Renewal Correspondence., (ML072360026)
08/29/2007	Meeting Agenda, Meeting Notice, from D. J. Ashley to Louise A. Lund, 09/27/2007, Forthcoming Meeting Public Meeting to Discuss Environmental Scoping for Vogtle Electric Generating Plant, Units 1 and 2 License Renewal Application Review., (ML072010123)
08/31/2007	Inspection Plan, from Charles Casto, NRC/RGN-II to Tom E. Tynan, Southern Nuclear Operating Co, Inc, Vogtle, Inspection Plan Report 22, for 09/01/2007 - 03/31/2009., (ML072430522)
09/05/2007	Letter from J. P. Leous, NRC to Tom E. Tynan, Southern Nuclear Operating Co, Inc, Environmental Site audit Regarding Vogtle Electric Generating Plant, Units 1 and 2, License Renewal Application (TAC Nos. MD5905 and MD5906).; (ML072400136)
09/19/2007	Memoranda, from K. C. Chang to Louise A. Lund, audit and Review Plan for Plant Aging Management Programs, Aging Management Reviews, and Time-Limited Aging Analyses for the Vogtle Electric Generating Plant, Units 1 and 2, (TAC Nos. MD5903 and MD5904)., (ML072560073)
09/20/2007	Press Release-II-07-047 - NRC to Solicit Public Input Sept. 27 in Waynesboro as Part of Vogtle Nuclear Plant License Renewal Application Review., (ML072630434)
10/01/2007	Meeting Agenda Vogtle Scoping Meeting Handouts (ML072750264)
10/12/2007	Letter from George B J, Southern Nuclear Operating Co, Inc. to NRC Document Control Desk, Vogtle Electric Generating Plant, Revision 14 to the Updated Final Safety Analysis Report, 10 CFR 50.59 Report, and Technical Specification Bases Changes., (ML072970843)
10/12/2007	Updated Final Safety Analysis Report (UFSAR), from Southern Nuclear Operating Co, Inc, to NRC Document Control Desk, Vogtle, Units 1 and 2, UFSAR Revision 14, Section 10.0, Steam and Power Conversion System through End., (ML072970857)
10/24/2007	Letter Request for Additional Information (RAI), from J. P. Leous, NRC to Tom E. Tynan, Southern Nuclear Operating Co, Inc, Request for Additional Information Regarding Severe Accident Mitigation Alternatives for Vogtle Electric Generating Plant, Units 1 and 2, License Renewal (TAC Nos. MD5905 and 5906), (ML072841107)

APPENDIX B: CHRONOLOGY	
Date	**Subject**
11/05/2007	Letter from Tom E. Tynan, Southern Nuclear Operating Co, Inc, to NRC Document Control Desk, Vogtle, Units 1 and 2 - Transmittal of License Renewal Application Environmental Site audit Information Request., (ML073240371)
12/10/2007	Memoranda, Request for Additional Information (RAI), from Rajender Auluck to Louise A. Lund, Request for Additional Information Regarding the Southern Nuclear Operating Company, Inc., License Renewal Application for the Vogtle Electric Generating Plant, Dated June 29, 2007., (ML073370344)
12/11/2007	Memoranda, Request for Additional Information (RAI), from Klein A, to Louise A. Lund, Request for Additional Information License Renewal Application Vogtle Electric Generating Plant Units 1 and 2 TAC No. MD5903 and MD5904., (ML073440088)
12/11/2007	Memoranda, Request for Additional Information (RAI), from Cranston G V/ADES/DSS, to Louise A. Lund, RAI Regarding the Vogtle Electric Generating Plant, Units 1 and 2, License Renewal Application, (ML073450435)
12/14/2007	Memoranda, Request for Access Authorization, from Harrison D G/ADES/DSS/SBPB, to Louise A. Lund, Request for Additional Information, Balance of Plant - Vogtle Steam Electric Station License Renewal Application (TAC NOS. MD5903 and MD5904)., (ML073400500)
12/20/2007	Letter License-Application for Facility Operating License (Amend/Renewal) DKT 50, Request for Additional Information (RAI), from Tom E. Tynan, Southern Nuclear Operating Co, Inc, to NRC Document Control Desk, Vogtle - Request for Additional Information Regarding the Analysis of Severe Accident Mitigation Alternatives., (ML073580627)
12/26/2007	Memoranda, Request for Additional Information (RAI), from Robert Dennig to Louise A. Lund, Vogtle License Renewal, Request for Additional Information (TAC MD5903 and MD5904)., (ML073520031)
12/27/2007	Memoranda, from Alan Hiser to Louise A. Lund, Vogtle Electric Generating Plants Units 1 and 2 - Request for Additional Information Regarding the Time Limiting Aging Analysis for the Steam Generator Tube Loss of Material (TAC MD5903 AND MD5904). , (ML073610598)
01/17/2008	Memoranda, Request for Additional Information (RAI), from Kamal Manoly to Louise A. Lund, Request for Additional Information Related to Section 2.4, "Scoping and Screening Results: Structures," and Section 4.5, "Concrete Containment Tendon Prestress" of Vogtle Electric Generating Plant Units 1 and 2 License Renewal Application., (ML080170546)
01/22/2008	Letter Request for Additional Information (RAI), from D. J. Ashley, NRC to Tom E. Tynan, Southern Nuclear Operating Co, Inc, Request for Additional Info for the Review of the Vogtle Electric Generating Plant, Units 1 and 2, LRA., (ML080080406)
01/23/2008	Memoranda, Request for Additional Information (RAI), from Terence Chan to Louise A. Lund, Request for Additional Information for Vogtle, Units 1 and 2, License Renewal Application (TAC Nos. MD5903 and MD5904)., (ML080250193)

APPENDIX B: CHRONOLOGY	
Date	**Subject**
01/28/2008	Letter Request for Additional Information (RAI), from D. J. Ashley, NRC to Tom E. Tynan, Southern Nuclear Operating Co, Inc, Request for Additional Information for the Review of the Vogtle Electric Generating Plant, Units 1 and 2, License Renewal Application., (ML080080416)
01/28/2008	Letter Meeting Summary, from D. J. Ashley, NRC, Summary of Telephone Conference Call Held on January 16, 2008, Between the US Nuclear Regulatory Commission and Southern Nuclear Operating Company, Inc., Concerning Request for Additional Information Pertaining to the Vogtle Units 1 & 2, License Renewal., (ML080170368)
01/28/2008	Letter Request for Additional Information (RAI), from D. J. Ashley, NRC to Tom E. Tynan, Southern Nuclear Operating Co, Inc, Request for Additional information for the Review of the Vogtle Electric Generating Plant, Units 1 and 2, License Renewal Application, Sections 2.4, 4.2, and 4.5., (ML080170653)
02/01/2008	Letter from Tom E. Tynan, Southern Nuclear Operating Co, Inc, to NRC Document Control Desk, Vogtle, Units 1 and 2 - License Renewal Application, Follow Up to Severe Accident Mitigation Alternatives Request for Additional Information Review Questions., (ML080360158)
02/08/2008	Letter from Tom E. Tynan, Southern Nuclear Operating Co, Inc, to NRC Document Control Desk, Vogtle, Units 1 & 2 - License Renewal - audit Question Responses., (ML080430373)
02/11/2008	Note to File including Telecon Record, from D. J. Ashley, NRC, Summary of Teleconference Between the NRC and Southern Nuclear Operating Company, Inc., Concerning Request for Additional Information Pertaining to Vogtle, Units 1 and 2, License Renewal Application., (ML080250159)
02/12/2008	Memoranda, Request for Additional Information (RAI), from M. A. Mitchell to Louise A. Lund, Request for Additional Information on Time Limited Aging Analyses of Reactor Vessels and Internals, Vogtle Nuclear Plant, Units 1 and 2 License Renewal Application (TAC Nos. MD5903 and MD5904)., (ML080440396)
02/21/2008	Letter from Tom E. Tynan, Southern Nuclear Operating Co, Inc, to NRC Document Control Desk, Vogtle, Units 1 and 2, License Renewal - Responses to 01/22/2008 RAIs., (ML080560267)
02/27/2008	Letter Request for Additional Information (RAI), from D. J. Ashley, NRC to Tom E. Tynan, Southern Co, Generation, Request for Additional Information for the Review of the Vogtle Electric Generating Plant, Units 1 and 2, License Renewal Application Section 4.2., (ML080510365)
02/27/2008	Letter from Tom E. Tynan, Southern Nuclear Operating Co, Inc, to NRC Document Control Desk, Vogtle, Units 1 & 2, Response to 01/28/2008 Request for Additional Information., (ML080600405)
03/17/2008	Memoranda, from Rajender Auluck to Louise A. Lund, Scoping and Screening Methodology audit Trip Report for the Southern Nuclear Operating Company, Inc., License Renewal Application for the Vogtle Electric Generating Plant, Units 1 and 2., (ML080640502)

\multicolumn{2}{c}{APPENDIX B: CHRONOLOGY}

Date	Subject
03/20/2008	Letter from Tom E. Tynan, Southern Nuclear Operating Co, Inc, to NRC Document Control Desk, Vogtle License Renewal Application Amendment 1 and Updated Future Action Commitment List., (ML080810440)
03/26/2008	Meeting Summary, from D. J. Ashley, NRC to Southern Nuclear Operating Co, Inc, Summary of Telephone Conference with Southern Nuclear Operating Company, Inc., Concerning RAI Pertaining to the Vogtle Electric Generating Plant, Unit 1 and 2, License Renewal Application., (ML080710453)
03/31/2008	Letter from Tom E. Tynan, Southern Nuclear Operating Co, Inc, to NRC Document Control Desk, Vogtle - Responses to 02/27/2008 Requests for Additional Information re: License Renewal., (ML080930299)
04/03/2008	Memoranda, Safety Evaluation, from Greg Cranston to Louise A. Lund, Safety Evaluation Regarding the Vogtle Electric Generating Plant, Units 1 and 2, License Renewal Application., (ML080940019)
04/18/2008	Memoranda, from George Wilson to Louise A. Lund, Vogtle Electric Generating Plant - Safety Evaluation Input - License Renewal Application, Section 2.5, TAC Nos. MD5903 and MD5904., (ML081090626)
04/22/2008	Letter Meeting Notice, from G. T. Hopper, NRC to Tom E. Tynan, Southern Nuclear Operating Co, Inc, 06/06/2008 Notice of Public Meeting with Southern Nuclear Co. On License Renewal Inspection Exit., (ML081130657)
04/28/2008	Letter Request for Additional Information (RAI), from D. J. Ashley, NRC to Tom E. Tynan, Southern Nuclear Operating Co, Inc, Vogtle, Units 1 & 2, Request for Additional Information, Review of License Renewal Application, Sections 3.3, 4.3 & 4.7.1., (ML081080189)
05/01/2008	Memoranda, Safety Evaluation, from Alan Hiser to Louise A. Lund, Safety Evaluation Input For Vogtle Electric Generating Plant Units 1 And 2 Regarding The Time Limiting Aging Analysis For Steam Generator Tube Loss Of Material (TAC NOS. MD5903 And MD5904)., (ML081221029)
05/05/2008	Memoranda, Safety Evaluation, from Terence Chan to Louise A. Lund, Safety Evaluation Input - Vogtle Electric Generating Plant Units 1 And 2, License Renewal Application., (ML081270730)
05/09/2008	Memoranda from Robert Dennig to Louise A. Lund, Safety Evaluation Report Input, License Renewal Application Vogtle Electric, Generating Plant, Unit 1 (TAC No. MD5903) and Unit 2 (MD5904)., (ML081190325)
05/13/2008	Meeting Summary, D. J. Ashley, NRC, Summary of Telephone Conference with Southern Nuclear Operating Company, Inc., Concerning Scoping and Screen RAI Pertaining to the Vogtle Electric Generating Plant, Unit 1 and 2, License Renewal Application., (ML081290027)
05/15/2008	Memoranda, Safety Evaluation Report, from Matthew Mitchell to Louise A. Lund, Vogtle Nuclear Plant, Units 1 and 2 License Renewal Application - Final Safety Evaluation Report Sections for Time Limited Aging Analyses (TAC NOS. MD5903 and MD5904)., (ML081360160)
05/15/2008	Memoranda, Safety Evaluation Report, Draft, from Matthew Mitchell to Louise A. Lund, Vogtle Nuclear Plant, Units 1 and 2 License Renewal Application - Final Safety Evaluation Report Section For The Aging Management Program For The Reactor Vessel (TAC NOS. MD5903 and MD5904)., (ML081360235)

APPENDIX B: CHRONOLOGY	
Date	**Subject**
05/16/2008	Memoranda, from Donald Harrison to Louise A. Lund, Transmittal of Draft Safety Evaluation for Vogtle Electric Generating Plant License Renewal Application (TAC Nos. MD5903 and MD5904)., (ML081300608)
05/29/2008	Letter from Tom E. Tynan, Southern Nuclear Operating Co, Inc, to NRC Document Control Desk, Vogtle, Units 1 and 2, License Renewal - Responses to 04/28/2008 RAIs., (ML081500825)
05/30/2008	Memoranda, Report, Miscellaneous, from Kamal Manoly to Louise A. Lund, Input to Safety Evaluation for Section 2.4, "Section 4.5," Concrete Containment Tendon Prestress: of Vogtle Electric Generating Plant Units 1 and 2 License Renewal Application., (ML081540013)
06/02/2008	Press Release-II-08-034: NRC Staff to Meet with Vogtle Nuclear Plant Officials to Discuss Nuclear Plant License Renewal Inspection., (ML081540316)
06/13/2008	Memoranda, from Rajender Auluck to Louise A. Lund, Scoping and Screening Draft SER Input Regarding the Southern Nuclear Operating Company, Inc., License Renewal Application for the Vogtle Electric Generating Plant, Units 1 and 2, dated June 29, 2007., (ML081280584)
06/23/2008	Letter from Tom E. Tynan, Southern Nuclear Operating Co, Inc, to NRC Document Control Desk, Vogtle, Units 1 and 2, License Renewal - Revised Requests for Additional Information (RAI) Responses., (ML081760126)
06/26/2008	Letter from Tom E. Tynan, Southern Nuclear Operating Co, Inc, to NRC Document Control Desk, Vogtle, Units 1 and 2, License Renewal - LRA Annual Update, LRA Amendment 2, and Updated Future Action Commitment List., (ML081930741)
07/17/2008	Letter from Tom E. Tynan, Southern Nuclear Operating Co, Inc, to NRC Document Control Desk, Vogtle, Units 1 and 2, License Renewal - Revised RAI Responses 3.3-1 & 3.4-1., (ML081990458)
07/18/2008	Inspection Report, Letter from K. M. Kennedy to Tom E. Tynan, Southern Nuclear Operating Co, Inc, IR 05000424-08-006 and 05000425-08-006; on May 19, 2008 - June 06, 2008; Vogtle Electric Generating Station, Unit 1 and 2; License Renewal Inspection Program, Aging Management Programs., (ML082000869)
08/11/2008	Letter from Tom E. Tynan, Southern Nuclear Operating Co, Inc, to NRC Document Control Desk, Vogtle, Units 1 and 2, License Renewal - Revised Updated Future Actions Commitment List, (ML082240518)
10/01/2008	Letter Audit Summary Report, from D. J. Ashley, NRC to Tom E. Tynan, Southern Nuclear Operating Co, Inc, Audit Summary report Regarding the License Renewal Application for the Vogtle Electric Generating Plant Units 1 and 2 (TAC No. MD5903) (ML082680295)
11/18/2008	Letter Request for Additional Information (RAI), from D. J. Ashley, NRC to Tom E. Tynan, Southern Nuclear Operating Co, Inc, Request for Additional information for the Review of the Vogtle Electric Generating Plant, Units 1 and 2, License Renewal Application, Sections 3.3 – Boral, (ML083170298)

APPENDIX B: CHRONOLOGY	
Date	**Subject**
11/19/2008	Letter from B. E. Holian, NRC to Tom E. Tynan, Southern Nuclear Operating Co, Inc, Transmitting the staff's Draft Safety Evaluation Report for the Review of the Vogtle Electric Generating Plant, Units 1 and 2, License Renewal Application, (ML082590218)
12/16/2008	Letter from Tom E. Tynan, Southern Nuclear Operating Co, Inc, to NRC Document Control Desk, Vogtle, Units 1 and 2, License Renewal – Response to 12/16/2008 Request For Additional Information - Boral, (ML083540900)
01/16/2009	Letter Meeting Summary, from D. J. Ashley, NRC, Summary of Telephone Conference Call Held on January 8, 2009, Between the US Nuclear Regulatory Commission and Southern Nuclear Operating Company, Inc., Concerning Request for Additional Information Pertaining to the Vogtle Units 1 & 2, License Renewal., (ML090090213)
01/20/2009	Letter, from Tynan T E, Southern Nuclear Operating Co, Inc, NRC/Document Control Desk NRC/NRR, "Vogtle Electric Generating Plant - License Renewal - LRA Amendment 4 and Updated Future Action Commitment List.," , (ML090210302)
01/26/2009	Memoranda Safety Evaluation, from Hiser A L, NRC/NRR/DCI/CSGB, to Pelton D L, NRC/NRR/DLR/RPB1, "Vogtle, Unit 1 and 2, Safety Evaluation Input License Renewal Application." , (ML090260689)
02/16/2009	Letter, from Tynan T E, Southern Nuclear Operating Co, Inc, NRC/Document Control Desk NRC/NRR, "Vogtle Electric Generating Plant - License Renewal - LRA Amendment 5 and Updated Future Action Commitment List.," , (ML090480128)
04/10/2009	Letter, from Mario V Bonaca, ACRS to Dale E. Klein, Chairman NRC, "Report On The Safety Aspects Of The License Renewal Application For The Vogtle Electric Generating Plant, Units 1 And 2", (ML090930450)

APPENDIX C

PRINCIPAL CONTRIBUTORS

This appendix lists the principal contributors for the development of this safety evaluation report (SER) and their areas of responsibility.

APPENDIX C: PRINCIPAL CONTRIBUTORS	
Name	Responsibility
E. Andruszkiewicz	Reviewer-Reactor Vessel
H. Ashar	Structural Engineering
D. Ashley	Project Manger
R. Auluck	Management Oversight
U. Baig	Reviewer-Mechanical
J. Beetle	Reviewer-Mechanical
B. Boger	Management Oversight
C. Brown	RII Inspection
T. Chan	Management Oversight
Y. Cheung	Reviewer-Metal Fatigue
J. Collins	Reviewer-Piping
G. Cranston	Management Oversight
J. Davis	GALL audit and Review
R. Dennig	Management Oversight
T. Ellis	SER Support
F. Farzam	Reviewer-Structures
Q. Gan	Reviewer- Mechanical TLAA
K. Green	SER Support
D. Harrison	Management Oversight
B.Heida	Reviewer-Mechanical
A. Hiser	Management Oversight
D. Hoang	GALL audit and Report
B. Holian	Management Oversight
N. Iqbal	Mechanical Engineering

APPENDIX C: PRINCIPAL CONTRIBUTORS	
Name	Responsibility
D. Jackson	Reviewer-Systems
A. Johnson	Reviewer-Steam Generators
C. Julian	Region II Inspection
L. Lake	Region II Inspection
L. Lund	Management Oversight
K. Manoly	Management Oversight
R. Mathew	GALL audit and Review
M. Mitchell	Management Oversight
J. Raval	Reviewer-Mechanical
S. Ray	Reviewer-Electrical
B. Rogers	Reviewer-Scoping
F. Saba	Reviewer-Mechanical
E. Smith	Reviewer-Scoping
G. Wilson	Management Oversight
Z. Xi	Reviewer-Civil/Structural

CONTRACTORS

CONTRACTOR	TECHNICAL AREA
Information Systems Labs	Plant Level Scoping Support

APPENDIX D

REFERENCES

This appendix lists the references used throughout this safety evaluation report (SER) for review of the license renewal application (LRA) for Vogtle Electric Generating Plant, Units 1 and 2.

APPENDIX D REFERENCES
NUREG-1800, Revision 1, "Standard Review Plan for Review of License Renewal Applications for Nuclear Power Plants," September 2005.
NUREG-1801, Revision 1, "Generic Aging Lessons Learned (GALL) Report, "September 2005.
NEI 95-10, Revision 6, "Industry Guideline for Implementing the Requirements of 10 CFR Part 54 - The License Renewal Rule," June 2005.
NUREG-1437, "Generic Environmental Impact Statement for License Renewal of Nuclear Plants (GEIS).
NUREG-1137, "Safety Evaluation Report Related to the Operation of Vogtle Electric Generating Plant, Units 1 and 2," through Supplement 9
American National Standards Institute (ANSI) B30.2-1976, "Overhead and Gantry Cranes"
NUREG-0612, "Control of Heavy Loads at Nuclear Power Plants"
NRC Bulletin 1987-01, "Thinning of Pipe Walls in Nuclear Power Plants, " November 1987
NRC Generic Letter 89-08 "Erosion/Corrosion-Induced Pip Wall Thinning," May 1999.
NRC Bulletin 2004-01, "Inspection of Alloy 82/182/600 Materials Used in the Fabrication of Pressurizer Penetrations and Steam Space Piping Connections at Pressurized-Water Reactors," May 2004.
NRC Bulletin 20003-02, "Leakage from Reactor Pressure Vessel Lower Head Penetrations and Reactor Coolant Pressure Boundary Integrity," August 2003.
NRC Order EA-03-009, "Issuance of Order Establishing Interim Inspection Requirements for Reactor Pressure Vessel Heads at Pressurized Water Reactors," February 2003.
NRC Generic Letter 88-05, "Boric Acid Corrosion of Carbon Steel Reactor Pressure Boundary Components in PWR Plants," March 1988.
NRC Generic Letter 89-13, "Service Water System Problems Affecting Safety-Related Equipment."

NRC FORM 335
(9-2004)
NRCMD 3.7

U.S. NUCLEAR REGULATORY COMMISSION

1. REPORT NUMBER
(Assigned by NRC, Add Vol., Supp., Rev.,
and Addendum Numbers, If any.)

BIBLIOGRAPHIC DATA SHEET

(See instructions on the reverse)

NUREG-1920, Volume 2

2. TITLE AND SUBTITLE	3. DATE REPORT PUBLISHED	
Safety Evaluation Report Related to the License Renewal of the Vogtle Electric Generating Plant	MONTH	YEAR
	April	2009
	4. FIN OR GRANT NUMBER	

5. AUTHOR(S)	6. TYPE OF REPORT
Donnie J. Ashley	Technical
	7. PERIOD COVERED *(Inclusive Dates)*
	06/27/2007 - 04/30/2009

8. PERFORMING ORGANIZATION - NAME AND ADDRESS *(If NRC, provide Division, Office or Region, U.S. Nuclear Regulatory Commission, and mailing address; if contractor, provide name and mailing address.)*

Division of License Renewal

Office of Nuclear Reactor Regulation

U. S. Nuclear Regulatory Commission

Washington, DC 20555-0001

9. SPONSORING ORGANIZATION - NAME AND ADDRESS *(If NRC, type "Same as above"; if contractor, provide NRC Division, Office or Region, U.S. Nuclear Regulatory Commission, and mailing address.)*

Same as above

10. SUPPLEMENTARY NOTES

11. ABSTRACT *(200 words or less)*

This document is a safety evaluation report (SER) on the license renewal application (LRA) for Vogtle Electric Generating Plant (VEGP), Units 1 and 2, as filed by the Southern Nuclear Operating Company, Inc. (SNC or applicant). By letter dated June 27, 2007, SNC submitted it application to the U. S. Nuclear Regulatory Commission (NRC) for renewal of the VEGP operating licenses for an additional 20 years. The NRC staff (the staff) prepared this report to summarize the results of its safety review of the LRA for complicance with Title 10 Part 54, "Requirements for Renewal of Operating Licenses for Nuclear Power Plants," of the Code of Federal Regulations (10CFR Part 54). In its June 27, 2007, submission letter, the applicant requested renewal of the operating licenses issued under Section 103 (Operating License Nos. NPF-68 and NPF-81) of the Atomic Energy Act of 1954, as amended, for Units 1 and 2 for a period of 20 years beyond the current expiration date of January 16, 2027, for Unit 1, and February 9, 2029, for Unit 2. VEGP is located approximately 26 miles southeast of Augusta, Georgia. The NRC issued the operating licenses for Unit 1 on March 16, 1987, and on March 31, 1989, for Unit 2. Units 1 and 2 are a dry ambient containment pressurized water reactor design. Westinghouse Electric supplied the nuclear steam supply system. The current licensed power output of each unit is 3625 megawatt thermal with a gross electrical output of approximately 1250 megawatt electric. The updated final safety analysis report (UFSAR) shows details of the plant and the site. This SER summarizes the results of the staff's safety review of the LRA and describes the technical details considered in evaluating the safety aspects of the units' proposed operation for an additional 20 years beyond the term of the current operating licenses.

12. KEY WORDS/DESCRIPTORS *(List words or phrases that will assist researchers in locating the report.)*	13. AVAILABILITY STATEMENT
10 CFR 54, license renewal, Vogtle, scoping and screening, aging management, time-limited aging analysis, TLAA, safety evaluation report	unlimited
	14. SECURITY CLASSIFICATION
	(This Page) unclassified
	(This Report) unclassified
	15. NUMBER OF PAGES
	16. PRICE

Printed
on recycled
paper

NUREG-1920, Vol. 2

Safety Evaluation Report: Related to the License Renewal of
Vogtle Electric Generating Plant, Units 1 and 2

April 2009

UNITED STATES
NUCLEAR REGULATORY COMMISSION
WASHINGTON, DC 20555-0001

OFFICIAL BUSINESS